HEADQUARTERS
DEPARTMENT OF THE ARMY
Washington, DC, 15 May 2004

LONG RANGE SNIPER RIFLE CALIBER .50, M107

REPORTING OF ERRORS

You can help improve this publication. If you find any mistakes or if you know of a way to improve the procedures, please let us know. Submit your DA Form 2028 (Recommended Changes to Publications and Blank Forms), through the Internet, on the Army Electronic Product Support (AEPS) website. The Internet address is http://aeps.ria.army.mil. If you need a password, scroll down and click on "ACCESS REQUEST FORM". The DA Form 2028 is located in the ONLINE FORMS PROCESSING section of the AEPS. Fill out the form and click on SUBMIT. Using this form on the AEPS will enable us to respond quicker to your comments and better manage the DA Form 2028 program. You may also mail, fax or E-mail your letter or DA Form 2028 direct to: AMSTA-LC-CI Tech Pubs, TACOM-RI, 1 Rock Island Arsenal, Rock Island, IL 61299-7630. The e-mail address is TACOM-TECH-PUBS@ria.army.mil. The fax number is DSN 793-0726 or Commercial (309) 782-0726.

TM 9-1005-239-10

WARNING SUMMARY

This weapon generates harmful levels of noise when firing! Hearing protection must be worn when firing this weapon.

Treat every firearm as if it were loaded. Never accept or take over a firearm from another until you yourself have checked personally to see that it is unloaded or in a completely safe condition.

Always control the muzzle of the firearm, keeping it pointed in a safe direction. With the M107 rifle, there is some danger in standing to the side of the muzzle due to blow by of hot, expanding gasses from the muzzle brake. Never point or aim any firearm at anything you are not willing to shoot!

Buffer and Buffer Spring are under heavy spring tension. Use caution when removing.

The bolt does NOT automatically remain to the rear when the rifle or magazine is empty. After the rifle is unloaded, and with the charging handle to the rear, always physically check the chamber to make certain the rifle is empty to preclude injury from accidental discharge.

WARNING SUMMARY - Continued

If the spring-loaded cam is lifted too far during disassembly, the spring may lose tension. If this occurs, the weapon could malfunction, or fire when unlocked, with the potential for serious injury.

When disassembling spring loaded parts, point components away from face/eyes to avoid possible injury if parts fly free.

The .50 caliber saboted light armor penetrator (slap) round is specifically prohibited from being fired in the M107. Forcing it to chamber and fire may cause serious injury to personnel and damage to the rifle.

Ammunition that predates 1965 should not be fired through this weapon.

Rifle must NOT be fired without both the midlock and rear lock pins firmly in place. Serious injury or death may result.

Ensure that the weapon is unloaded and on 'SAFE' before performing any functional procedures.

WARNING SUMMARY - Continued

Never try to force a cartridge to chamber. If the bolt does not fully close, remove the magazine, clear the weapon, and check for obstructions, but do NOT attempt to fire. Serious injury or damage to the weapon may result.

This rifle must not be fired without the muzzle brake firmly in place on the barrel because the action can thus be overstressed, and critical parts breakage can occur.

CAUTION SUMMARY

Be sure the hook and bar are properly mated, or the rifle can be damaged by final assembly motion.

Double hearing protection should be worn when firing, since harmful levels of noise are generated.

Whenever inserting cleaning rods or other devices through the muzzle end of the barrel, be especially careful NOT to damage the muzzle's crown, since it could affect the accuracy of the rifle.

CAUTION SUMMARY CON'T

Do NOT leave rounds in the magazine for extended periods of time since this will cause the spring to lose tension and may cause a malfunction.

When removing the bolt carrier from the lower receiver, ensure carrier is completely forward of the housing before lifting to avoid damaging the lower receiver.

Do not allow bore cleaner or any solvent to remain in contact with the impact (barrel) bumpers as it may cause plastic composite material to deteriorate.

Do NOT allow sunlight to shine directly through the scope. Light focused on the crosshairs and mil dots might warp them.

Use extreme care to avoid scratching the lenses. Do NOT allow solvents to touch lenses.

INSERT LATEST CHANGED PAGES / WORK PACKAGES. DESTROY SUPERSEDED DATA.

LIST OF EFFECTIVE PAGES / WORK PACKAGES

Dates of issue for original and changed pages / work packages are:

Original .... 0 15 May 2004

TOTAL NUMBER OF PAGES FOR FRONT AND REAR MATTER IS 30 AND TOTAL NUMBER OF WORK PACKAGES IS 22 CONSISTING OF THE FOLLOWING:

Page / WP No.	*Change No.	Page / WP No.	*Change No.
Title	0	i – iv	0
a - d	0	WP 0001 00 - WP 0022 00	0
A	0	Index -1 - Index - 11	0
B Blank	0	Index - 12 blank	0

*Zero in this column indicates an original page or work package.

TABLE OF CONTENTS

TABLE OF CONTENTS – Continued

WP Sequence No.

HOW TO USE THIS MANUAL

The safest, easiest, and best way to operate and maintain the M107 Sniper Rifle is to use this manual. Learning to use this TM is as easy as reading through the next few pages of this section. Knowing what is in this manual and how to use it will save you time and work and will help you avoid exposing yourself to unnecessary hazards while performing your job.

So Where Do You Start?

Right here, if this is the first time you are using this TM. Be sure to completely read this section on how to use this manual first. There's a lot of information here that you need to know.

Organization

This manual covers the operation and maintenance of the M107 Sniper Rifle. The manual itself is divided into six chapters, including supporting information. The six chapters and what they contain, are found in the Table of Contents in the front of this manual. For example, to learn about operating the M107 Sniper Rifle, you would look in the table of contents and discover that Chapter 2 provides all pertinent information about the operation of the M107 Sniper Rifle. Since Chapter 2 covers a great deal of information, you will have to scan the chapter to find the specific information you will need.

In Chapter 6, you will find the supporting information. Each work package provides specific information that will assist you in performing the various operational and maintenance tasks. The work packages provide such information as additional References (i.e., other TMs or FMs), as in WP 0020 00, and Expendable and Durable Items List, as in WP 0022 00. Become familiar with all supporting information work packages and what they contain before beginning any operational or maintenance task.

Am I Ready To Use The TM?

If you've taken the time necessary to read this section, and are sure of the location and arrangement of the different sections of this TM, you are ready to begin. Remember, this TM has been arranged with you, the user, in mind. Your safety and ability to perform the operational and maintenance tasks in the most efficient manner possible hinge on your ability to perform and understand the information contained in this manual. If you fully understand the arrangement and purpose of this TM, and have taken the time to read through this section, you will have no trouble operating and maintaining the rifle in the manner for which it was designed.

TM 9-1005-239-10

CHAPTER 1

INTRODUCTION

OPERATOR
LONG RANGE SNIPER RIFLE, M107
NSN 1005-01-469-2133
GENERAL INFORMATION

SCOPE

Type of Manual

Operator's Manual.

Model Numbers and Equipment Names

Rifle, Caliber .50, Sniper with Day Optical Sight and Carrying Case, M107.

Purpose of Equipment

The M107 Long Range Sniper Rifle (LRSR) is a man-portable, direct line of sight weapon system capable of providing precision fire on targets at a distance of up to 1000 meters.

MAINTENANCE FORMS, RECORDS, AND REPORTS

Department of the Army forms and procedures used for equipment maintenance will be those prescribed by (as applicable) DA PAM 738-750, Functional Users Manual for the Army Maintenance Management System (TAMMS); DA PAM 738-751, Functional Users Manual for the Army Maintenance Management System; or AR 700-138, Army Logistics Readiness and Sustainability.

REPORTING EQUIPMENT IMPROVEMENT RECOMMENDATIONS (EIR)

If your M107 Sniper Rifle needs improvement, let us know. Send us an EIR. You, the user, are the only one who can tell us what you don't like about your equipment. Let us know why you don't like the design or performance. Put it on an SF 368 (Product Quality Deficiency Report). Mail it to the address specified in DA PAM 738-750, Functional Users Manual for the Army Maintenance Management System (TAMMS), or as specified by the acquiring activity. We will send you a reply.

CORROSION PREVENTION AND CONTROL (CPC)

Corrosion Prevention and Control (CPC) of army materiel is a continuing concern. It is important that any corrosion problems with this item be reported so that the problem can be corrected and improvements can be made to prevent the problem in future items.

While corrosion is typically associated with rusting of metals, it can also include deterioration of other materials, such as rubber and plastic. Unusual cracking, softening, swelling, or breaking of these materials may be a corrosion problem.

If a corrosion problem is identified, it can be reported using SF 368, Product Quality Deficiency Report. Use of Key words such as "corrosion", "rust", "deterioration", or "cracking" will ensure that the information is identified as a CPC problem.

The form should be submitted to the address specified in DA PAM 738-750, Functional Users Manual for the Army Maintenance Management System (TAMMS).

DESTRUCTION OF ARMY MATERIAL TO PREVENT ENEMY USE

Refer to TM 750-244-7.

PREPARATION FOR STORAGE OR SHIPMENT

Storage

WARNING

Do not store the weapon with live ammunition in either the chamber or magazine.

Disassemble the weapon into its three major assemblies (refer to WP 0013 00), apply a coat of CLP (item 1, WP 0022 00), and store the weapon in its carrying case.

PREPARATION FOR STORAGE AND SHIPMENT - Continued

Shipment

Ship weapons requiring depot level maintenance in accordance with the disposition instructions and pertinent auto-retrograde message.

WARNING

Under no circumstances should the weapon be shipped while it contains live ammunition, either in the shipping box, magazine, or chamber.

a. Ensure that the magazine and chamber are clear of ammunition (refer to WP 0005 00).

b. Complete forms in accordance with specifications and detail the required maintenance as thoroughly as possible.

c. Clean the weapon (refer to WP 0005 00).

d. Place the weapon, broken down into its three major groups, in a carrying case and place it in a shipping box.

e. Mark the box in accordance with MIL-STD-129.

NOMENCLATURE CROSS-REFERENCE LIST

Official Name	Common Name
Cartridge Magazine	Magazine Assembly
Quick Release Pin	Midlock Pin
Quick Release Pin	Rear Lock Pin

LIST OF ABBREVIATIONS/ACRONYMS

Abbreviation/Acronym	Name
API	Armor-Piercing Incendiary
APIT	Armor-Piercing Incendiary Tracer
CLP	Cleaner, Lubricant, and Preservative
LAW	Lubricant, Arctic Weather
LSA	Lubricant, Small Arms
LSAT	Lubricant, Small Arms (with Teflon)
LRSR	Long Range Sniper Rifle
MOA	Minute of Angle
RBC	Rifle Bore Cleaner

QUALITY OF MATERIAL

Material used for replacement, repair, or modification must meet the requirements of this manual. If quality of material requirements are not stated in this manual, the material must meet the requirements of the drawings, standards, specifications, or approved engineering change proposals applicable to the subject equipment.

SAFETY, CARE, AND HANDLING

Refer to WP 0019 00 for general ammunition safety, care, and handling.

END OF WORK PACKAGE

OPERATOR
LONG RANGE SNIPER RIFLE, M107
NSN 1005-01-469-2133
EQUIPMENT DESCRIPTION AND DATA

EQUIPMENT CHARACTERISTICS, CAPABILITIES, AND FEATURES

Capabilities

The M107 is a long-range sniper weapon system which utilizes standard caliber .50 ammunition. The M107 is a man-portable, direct line-of-sight system capable of providing precision fire on targets at distances up to 1000 meters.

Functional Description

The M107 is a semi-automatic, air-cooled, box magazine-fed rifle chambered for .50 caliber ammunition. This rifle operates by means of the short recoil principle, rather than gas.

Characteristics

The basic M107 rifle is equipped with bipod, muzzle brake, carrying handle, and 10-round removable magazine. The M107 system is composed of the rifle and a sniper scope, plus spare magazines. The rifle is also supplied with fitted carrying case, the requisite cleaning kit drag bag, cleaning equipment, and the telescope adjustment tools.

LOCATION AND DESCRIPTION OF MAJOR COMPONENTS

a. Upper Receiver Assembly. Includes the front sight, accessory base, carrying handle, muzzle brake, and barrel.

b. Rail. Used to attach the scope, the carrying handle, and accessory optic sights.

c. Carrying Handle. Steel stock with a hard plastic handle.

d. Front Sight. A 0.075 in. (0.19 cm) post with a protective, anti-glare ring.

e. Muzzle Brake. Critical to the functioning of the weapon; absorbs approximately 70 percent of the recoil.

f. Barrel. Length is 29 in. (73.7 cm) with eight lands and grooves in a uniform right-hand twist, one turn in 15 in. (38.1 cm). Muzzle end is threaded to accept a muzzle brake; breech end has a barrel extension integral to the locking function.

g. Bolt Assembly. Houses the firing pin, extractor, and ejector.

h. Bolt Carrier Assembly. Consists of the bolt, firing pin, all extraction and ejection mechanisms, cocking lever, and sear.

i. Bipod Assembly. Detachable forward support system composed of retractable legs and extending foot pads.

j. Lower Receiver Assembly. Includes detachable bipod assembly, buffer assembly, midlock pin, and trigger mechanism.

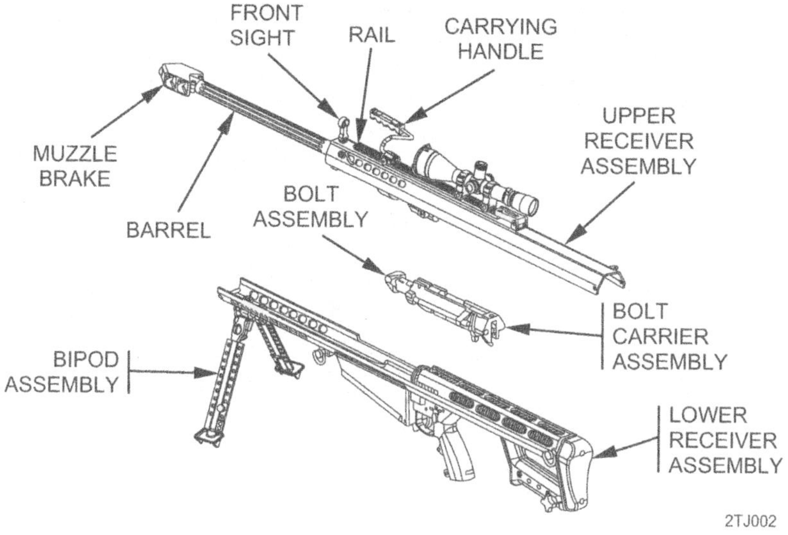

EQUIPMENT DATA

The M107 Rifle System comprises the rifle with a fixed variable 4.5X14 power sniper scope, and 6 magazines. The rifle is also supplied with a fitted dirt-tight and watertight carrying case, cleaning kit, and telescope adjustment tools.

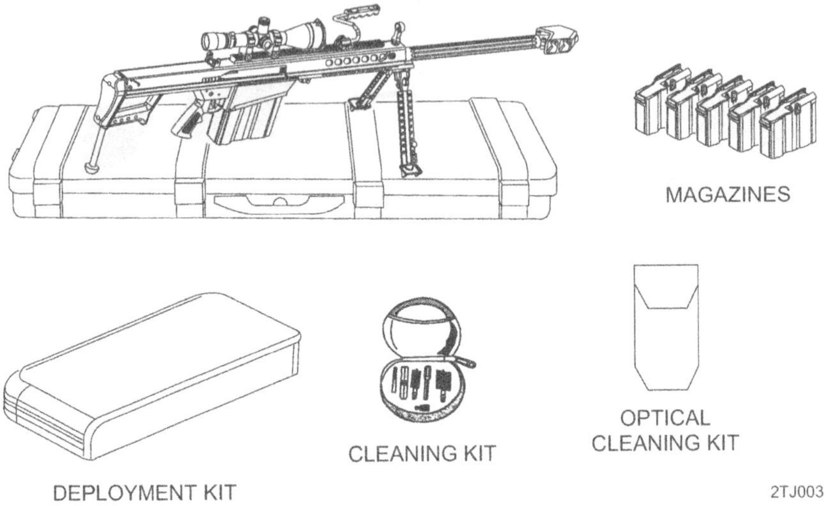

TM 9-1005-239-10 0002 00

EQUIPMENT DATA - Continued

Table 1. Specifications and Capabilities.

Specification	
Weight (gun and scope)	28.5 lbs. (12.9 kilograms) (unloaded)
Overall length assembled	0.57 in. (88.9 cm)
Length in takedown mode	38 in. (86.36 cm)
Barrel length	29 in. (73.66 cm)
Magazine capacity	10 rounds
Stock	Integral with lower receiver assembly (steel)
Safety	Manual thumb-lever
Muzzle velocity	Approx. 2,800 f.p.s. (853 meters/sec) (with standard 660 grain bullet)
Maximum range	Approx 7,450 yards (6,800m) (with Standard 660 grain bullet)
Maximum effective range	2,000 yards (1,830 meters) (with standard 660 grain bullet)
Muzzle energy	11,500 foot-pounds (15,582 joules)
Magazine weight	Ten rounds 4.12 lbs. Eight rounds 3.62 lbs.
Sight Type	Leupold 4.5X14 Vary X

Table 1. Specifications and Capabilities – Continued.

Specification	
Length	12 5/8 in.
Reticle	Mil dot
Lens	50 mm
Elevation	1 Click equals 1/4 MOA at 100 Meters
Windage	1 Click equals 1/4 MOA at 100 Meters
Eye Relief	3" – 6"
Compatible Ammunition	
	MK211 Mod 0, Caliber .50 API Cartridge M33, Caliber .50 Ball Cartridge M17, Caliber .50 Tracer Cartridge M8, Caliber .50 API Cartridge M20, Caliber .50 APIT Cartridge M1A1, Caliber .50 Blank Cartridge

END OF WORK PACKAGE

OPERATOR
LONG RANGE SNIPER RIFLE, M107
NSN 1005-01-469-2133
THEORY OF OPERATION

CYCLE OF OPERATION

The cycle of operation for the M107 Rifle is broken down into eight basic steps (more than one step may occur at the same time).

1. Feeding: The force of the mainspring pushes the bolt forward toward the barrel extension, stripping a cartridge from the magazine and loading it into the chamber (by hand when first loading, by semiautomatic action afterwards).

2. Chambering: The bolt forces the round fully into the firing chamber, and the extractor snaps over the case rim. Blockages (dirt or debris) can prevent full chambering, as can dirty, bent, dented, or otherwise faulty ammunition.

3. Locking: During chambering the bolt enters the barrel extension, and the bolt latch engages the bolt latch trip (inside-top of the upper receiver assembly, just behind barrel extension). The bolt latch is then depressed, allowing the bolt to retract into the bolt carrier. The bolt, in turn, rotates due to the cam slot and is locked when its three locking lugs rotate into place in the barrel extension, closing the firing chamber.

CYCLE OF OPERATION - Continued

4. Firing: Pulling the trigger pivots it on the trigger housing pin and presses on the transfer bar, causing the bar to rise. The transfer bar engages the sear (housed in the bolt carrier), forcing it upward and out of engagement with the firing pin extension. The firing pin extension, under spring power, forces the firing pin forward to strike the primer of the cartridge.

5. Unlocking: When the cartridge is fired, gas pressure exerts a thrust on the bolt face via the case head. The bolt carrier carries the bolt and barrel extension to the rear until the accelerator, protruding beneath the bolt carrier, contacts a shoulder in the trigger housing area. The accelerator is then pivoted up, causing the accelerator rod to be pushed out of the bolt carrier. As it protrudes from the front of the bolt carrier, it separates the bolt carrier from the barrel extension. Because of the cam slot in the side of the bolt, the bolt rotates as it is pulled and unlocks from the barrel extension.

6. Cocking: As the bolt recoils to the rear, the cocking lever "rides" the transfer bar back and down, causing it to disconnect from the trigger. The transfer bar is then held down in this position by the disconnector and is not released until pressure is released from the trigger. After disconnection, the cocking lever swings on its pin and overrides the transfer bar. The other end of the cocking lever protrudes into the bolt carrier and into the firing pin extension. As the cocking lever pivots, it withdraws the firing pin and compresses the firing pin extension sprang. The firing pin extension then catches the sear.

7. Extraction: As the bolt-locking lugs rotate away from the barrel extension, the bolt withdraws from the barrel and the bolt latch locks the bolt in its extended position. The extractor located on the bolt face and hooked over the rim of the fired case, pulls the case from the firing chamber.

8. Ejection: As soon as the fired case has been extracted and has cleared the rear of the barrel extension, it is expelled from the rifle by the spring-powered ejector.

END OF WORK PACKAGE

CHAPTER 2

OPERATOR INSTRUCTIONS FOR LONG RANGE SNIPER RIFLE, M107

OPERATOR
LONG RANGE SNIPER RIFLE, M107
NSN 1005-01-469-2133
DESCRIPTION AND USE OF OPERATOR CONTROLS

GENERAL

The following paragraph contains an illustration that shows the location of each control for the M107 Long Range Sniper Rifle (LRSR). The numbers on the illustration are keyed to the tabular listing which contains the name and functional description of the controls.

CONTROLS

Table 1 describes the controls for the sniper rifle. The key number in the table corresponds with the callout number in the illustration.

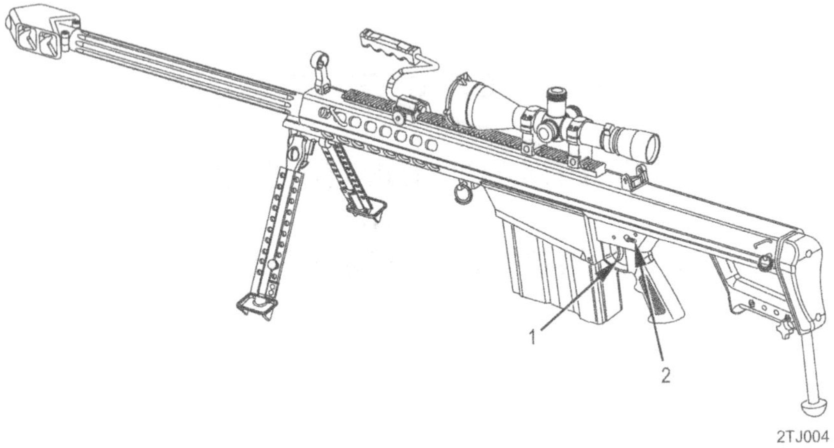

Table 1. M107 Sniper Rifle Controls.

KEY	CONTROL	FUNCTION OR USE
1	Trigger	Fires the weapon by squeezing toward the rear of the rifle.
2	Safety Switch	Prevents and allows firing of the rifle.

END OF WORK PACKAGE

OPERATOR
LONG RANGE SNIPER RIFLE, M107
NSN 1005-01-469-2133
OPERATION UNDER USUAL CONDITIONS

INITIAL SETUP:

Tools and Special Tools
5 Piece Cleaning Rod (item 10, WP 0021 00)

LOADING THE MAGAZINE

1. Center a cartridge (1) between the feed lips (2) in the magazine (3) and press down until the cartridge snaps under the lips.

CAUTION

The magazine has a ten round capacity. The preferred load is 8 to 9 rounds. Load no more than 9 rounds or damage will occur.

2. Ensure the cartridges are pushed all the way to the rear and that there is no interference between the magazine (3) and the nose of the cartridges.

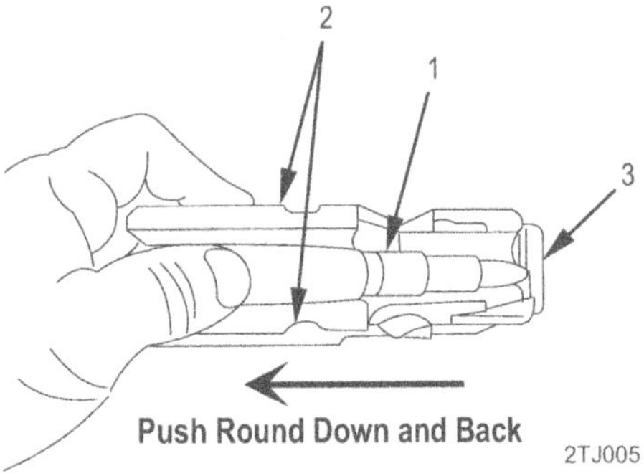

INSERTING THE MAGAZINE INTO WEAPON

WARNING

Ensure the safety is in the "safe" position and that the chamber is clear of any obstruction.

1. Prior to inserting the magazine (1), grasp the charging handle (2) and dry-cycle the weapon several times (work the bolt all the way back and forth). If any damage has occurred to either the upper or the lower receiver assemblies during shipping, the bolt carrier will not move freely.

2. Insert the magazine (1) into the magazine well of the lower receiver assembly, tilting the magazine so that the bullets point upward. Place the magazine hook (3), located on the front of the magazine, onto the hinge, located in the front of the magazine well. Now rotate the magazine upwards until you hear the click of it locking into the rear magazine catch (4).

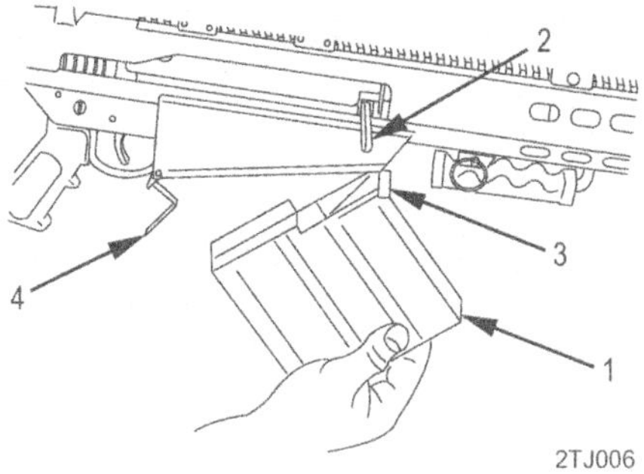

INSERTING THE MAGAZINE INTO WEAPON - Continued

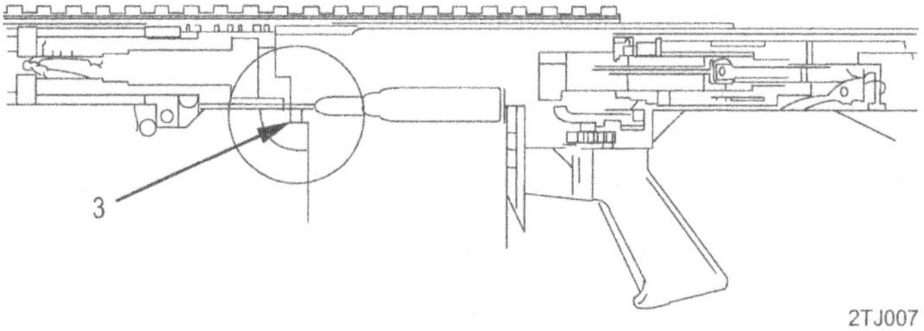

NOTE

Ensure the magazine hook (3) is in the correct space. Tugging on the magazine is a good check for proper seating.

CHAMBERING A CARTRIDGE

WARNING

Ensure the safety is in the "safe" position and that the chamber is clear of any obstruction.

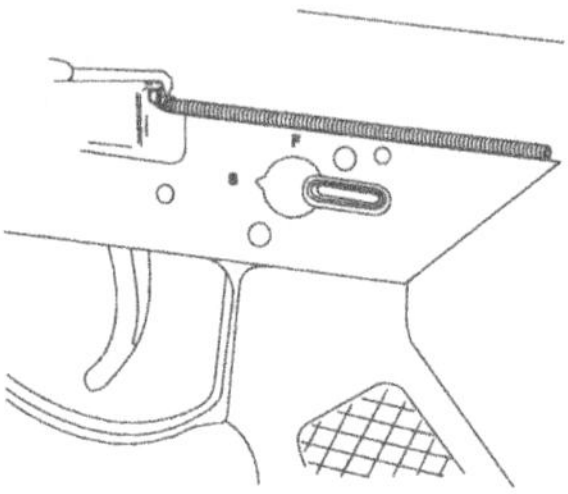

2TJ008

CHAMBERING A CARTRIDGE - Continued

WARNING

Never try to force a cartridge into the chamber. If the bolt does not fully close, clear any obstructions until a round is successfully chambered. Never attempt to fire a round if the bolt does not fully close.

1. With the muzzle pointed down range and in a safe direction, move the bolt handle up and all the way to the rear. Release the bolt handle chambering the first round from the magazine. Tap the bolt handle to ensure the bolt is fully forward.

WARNING

Because the rifle is recoil-operated, the shooter must be positioned squarely behind the weapon, with the recoil pad firmly against the shoulder. Anything less may result in injury, discomfort, or failure of the action to cycle correctly.

2. Shooter is positioned squarely behind the rifle. Rotate safety selector switch to the "FIRE" position. Ensure shoulder pad is properly positioned on shoulder before squeezing the trigger. The rifle will fire one round for each squeeze of the trigger until the magazine and chamber are empty.

3. With the safety in the "safe" position and the muzzle pointed in a safe direction, pull the charging handle to the rear until it stops, then release it. The rifle then loads and locks under its own spring power.

Jamming

CAUTION

Whenever it is necessary to insert cleaning rods (item 10, WP 0021 00) or other devices through the muzzle end of the barrel, be especially careful not to damage the muzzle's crown, since it could affect the accuracy of the rifle.

If a cartridge or round jams in the chamber, use a cleaning rod to carefully push it out.

UNLOADING THE RIFLE

WARNING

The bolt does not automatically remain to the rear when the rifle or magazine is empty. After the rifle is unloaded, and with the charging handle to the rear, always physically check the chamber to ensure that the rifle is empty.

UNLOADING THE RIFLE - Continued

1. Place the weapon safety on "safe". Press the magazine catch (1) forward towards the magazine (2), and remove the magazine.

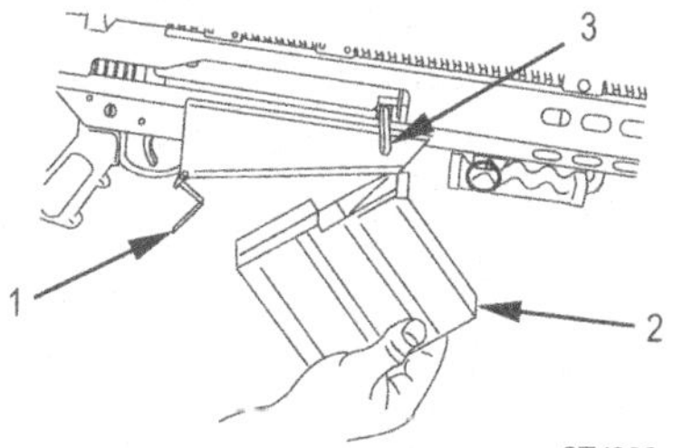

2TJ009

NOTE

Remember that the bolt does NOT automatically remain to the rear when the rifle or magazine is empty.

2. Pull the charging handle (3) to the rear, which will eject any cartridge still chambered.

3. After the rifle is unloaded, and with the charging handle (3) to the rear, look into the chamber make certain the breech area and chamber are empty.

Clearing

CAUTION

Whenever it is necessary to insert cleaning rods or other devices through the muzzle end of the barrel, be especially careful not to damage the muzzle's crown, since it could affect the accuracy of the rifle.

CAUTION

Do not leave rounds in the magazine for extended periods of time since this will cause the spring to lose tension and may cause a malfunction.

UNLOADING THE MAGAZINE

CAUTION

Only remove ammunition from the magazine with your hands.

Hold the magazine (1) with one hand with the cartridge tips (2) facing away from you. Push each of the cartridges out of the magazine until all are ejected.

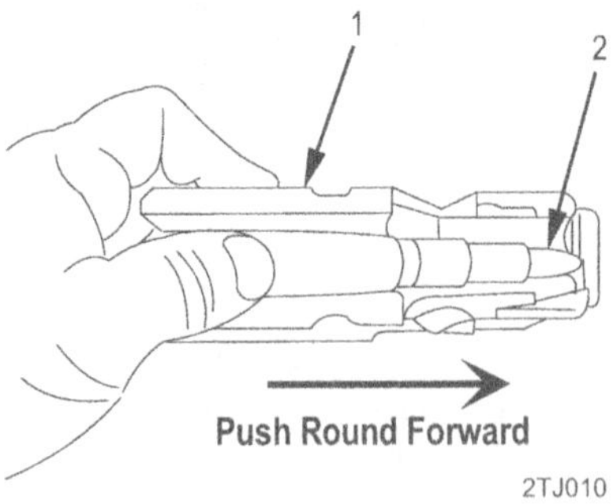

END OF WORK PACKAGE

OPERATOR
LONG RANGE SNIPER RIFLE, M107
NSN 1005-01-469-2133
SIGHTING SYSTEMS

ZEROING THE IRON SIGHT

1. Assume a prone-supported firing position 100 meters from the zero target (NSN 6920-00-900-8205). Align index line with the 100-meter range line on the elevation scale (1).

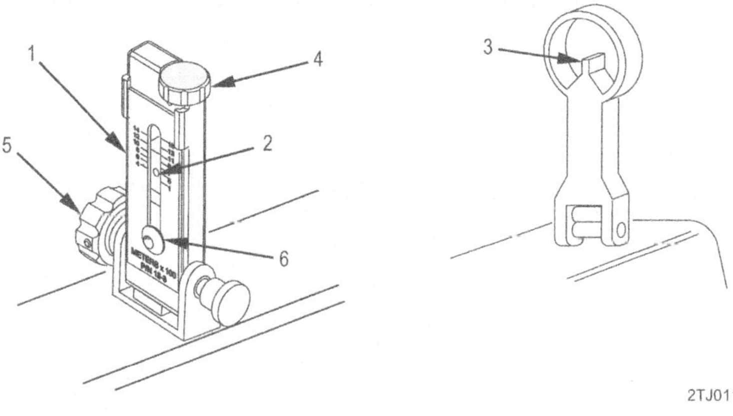

2TJ011

2. Align windage index line with windage zero index line on base of iron sight. Align the rear peep sight (2) with the front sight post (3).

NOTE

One click of elevation knob (4) moves the strike of the round 1.6 in. (4.06 cm) at 100 meters.

One click of windage knob (5) moves the strike of the round 0.75 in. (1.90 cm) at 100 meters.

3. Obtain the proper sight picture by aligning the rear peep sight (2) with front sight post (3). Fire a three round shot group at center mass of the zero target, maintaining the same aim point with each shot. Note the strike of the rounds and make windage and elevation adjustment accordingly.

4. Continue firing three round shot groups, making windage and elevation adjustments as necessary until the shot group is center mass on the zero target.

5. Once the shot group is center mass, loosen the screw (6) on the elevation scale (1). Slide the elevation scale until the 100 meter range index line is aligned with the index line on the rear peep sight (2), tighten the screw.

6. Confirm zero by setting the elevation scale (1) to the 500-meter line and fire a three round shot group at a 500-meter target.

DAYLIGHT SCOPE

There are seven basic parts to the daylight scope, the objective lens (front lens) (1), the elevation adjustment turret (2), the windage adjustment turret (3), the power selector ring (4), the eyepiece lock ring (5), parallax adjustment turret (6), and the eyepiece lens (7).

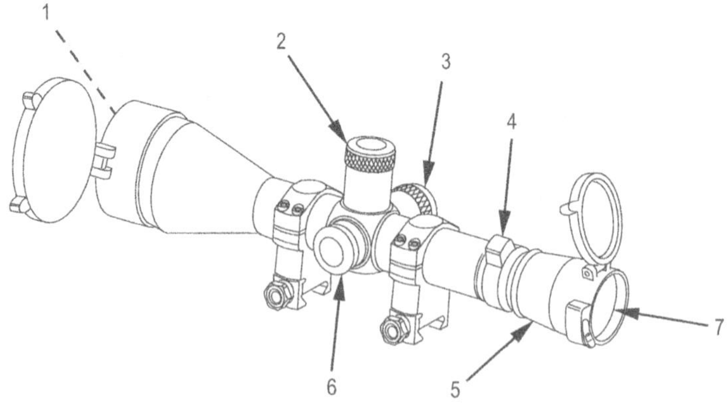

2TJ012

Live Fire Zero

1. Zero the scope on a known-distance range on a zero target (NSN 6920-00-900-8205).

2. Assume a prone-supported firing position 100 meters from the target. Turn the elevation adjustment turret (1) clockwise until it bottoms out. Once at the bottom, turn the elevation adjustment turret counterclockwise 34 minutes or 136 clicks. Turn the windage adjustment turret (2) clockwise until it bottoms out. Once at the bottom, turn the windage adjustment turret counterclockwise 52 minutes or 208 clicks.

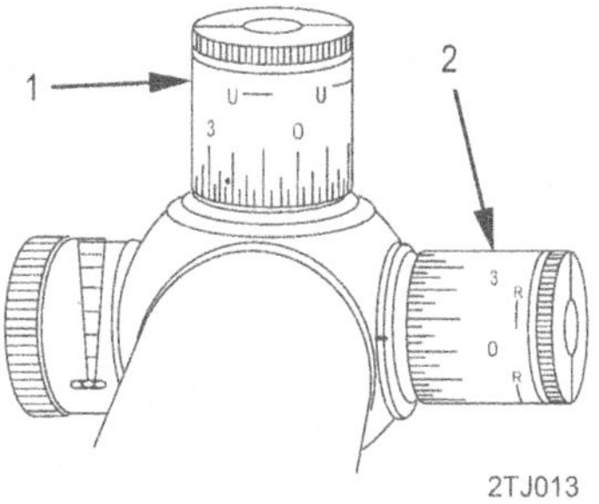

2TJ013

DAYLIGHT SCOPE - Continued

3. Fire three rounds at the center of the target, keeping the same aiming point each time and triangulate. Note the strike of the rounds and make windage and elevation adjustment accordingly. Continue firing three round shot groups making windage and elevation adjustments as necessary until the shot group is center mass on the zero target.

NOTE

Each click of windage or elevation equals 1/4 minute of angle (MOA). Therefore at 100 meters, 1 MOA equals approximately 1 inch.

4. Once the shot group is centered, loosen the screws on the elevation (3) and windage adjustment dials (4). Turn the elevation adjustment turret (1) to the index line marked "1" and tighten screws. Turn the windage adjustment turret (2) until the "0" on the windage adjustment turret is lined up with the windage index line and tighten screws.

5. After zeroing at 100 meters, confirm by setting the elevation adjustment turret (1) to 500 meters and firing a three round shot group at a 500 meter target.

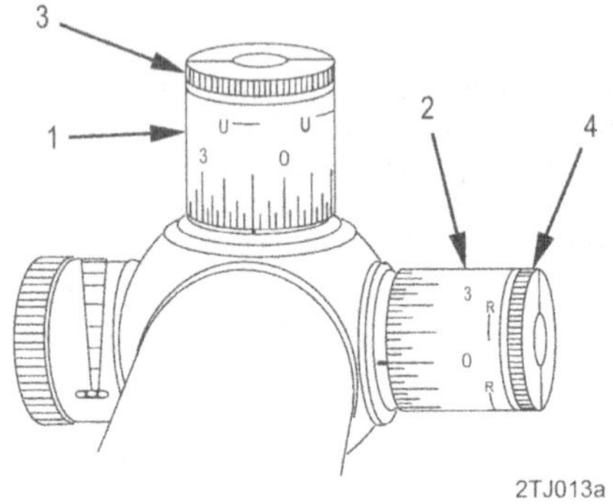

TM 9-1005-239-10 0006 00

DAYLIGHT SCOPE - Continued

Table 1. Estimates for Zeroing the Scope.

Range	Clicks	Minute
100 (Zero)	136*	34*
200	26	6.5
300	7	1.75
400	7	1.75
500	6	1.5
600	9	2.25
700	12	3
800	16	4
900	18	4.5
1000	24	6
*From the bottom		

The above are estimates which may require individual situation changes.

Eye Relief

WARNING

If the scope is mounted too far to the rear, the eyepiece can injure the shooter's brow. Shooting at an uphill angle also increases this hazard.

NOTE

The daylight scope is manufacturer-mounted with the rear ring mount on slot 4 of the rail, torqued to 65 in. lb. Always ensure the scope is mounted with a torque value of 65 in. lb. Remove the ring mount from the rail using a 1/2 in. wrench turning counterclockwise. Tighten the ring mount nuts using only the preset 65 in. lb. T-handle torque wrench.

Eye relief is always obtained from the prone position.

DAYLIGHT SCOPE - Continued

1. Mount the scope (1) with the rear ring mount (2) on slot 4 of the rail (3). Using T-handle torque wrench, torque ring mount nuts (4) to 65 in. lb. Hold the rifle in normal firing position. Set the scope to the highest possible magnification.

2. Slowly move the scope (1) forward or rearward on the rail (3) until a full field of view is seen (if necessary). Position scope here for rough eye relief. To achieve optimum eye relief, loosen the optic rings and slide the telescope tube forward or backward.

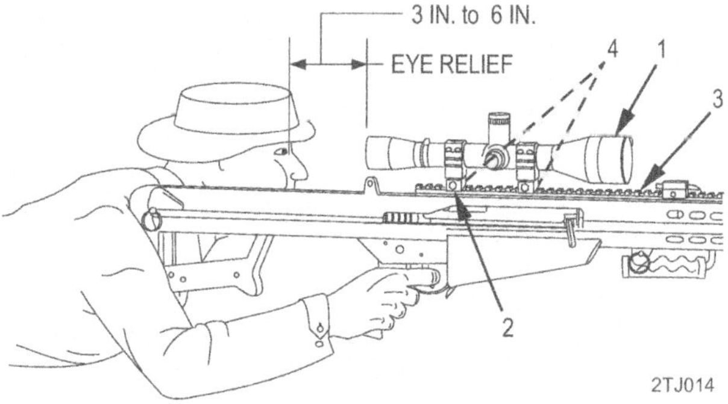

NOTE

Remember that aiming uphill reduces eye relief.

3. Without disturbing the eye relief position, rotate the scope until the elevation adjustment dial is on the top of the scope.

4. From a firing position, check to be sure that the vertical hair of the reticle aligns with the vertical axis of the firearm. Misalignment will not affect the accuracy at short to moderate distances but it can diminish long range accuracy.

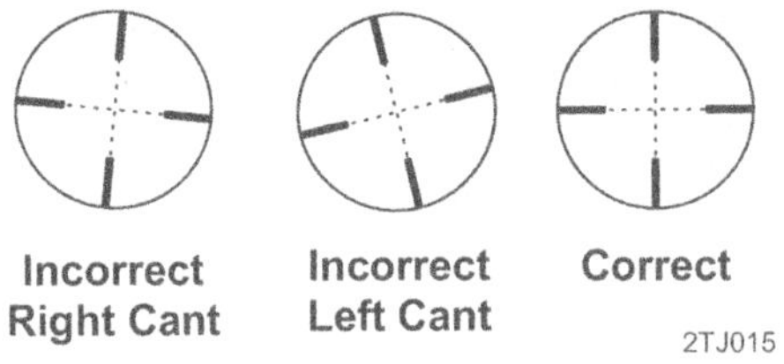

TM 9-1005-239-10 0006 00

DAYLIGHT SCOPE - Continued

5. When satisfied with eye relief, tighten the ring screws evenly and securely. Start in one corner and tighten a small amount, then tighten the screw in the opposite corner, then the screw above/below then across corner. Continue the pattern until all screws are tightened. This ensures an even tightening and prevents twisting of the scope.

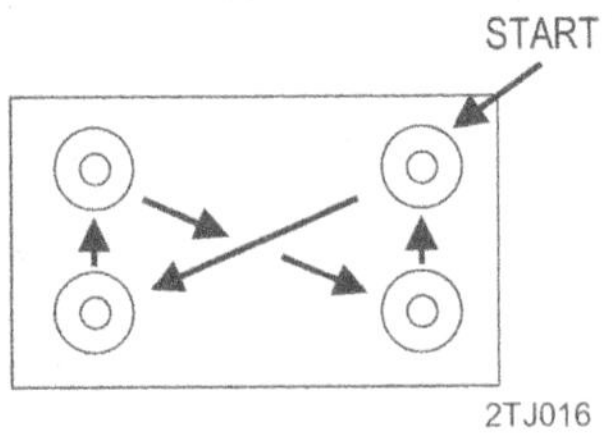

Focusing The Reticle

1. Point the scope at a light colored background object. With established eye relief, the reticle should appear sharp and crisp. If it does not, adjust the focus by means of the eyepiece.

2. Turn the eyepiece until the reticle appears very fuzzy then turn eyepiece until the reticle appears clear and sharp.

3. When satisfied with the image of the reticle, turn the lock ring so that it rests firmly against the eyepiece.

Parallax

CAUTION

Do NOT lubricate the power selector ring.

DAYLIGHT SCOPE - Continued

NOTE

Reticle should be focused before turning the side focus adjustment dial. If it is NOT, follow instructions for focusing the reticle.

Parallax is the apparent movement of the target relative to the reticle when eye is moved away from the center point of the eyepiece.

No numbers indicating distance appear on the dial as all adjustment is judged by the image itself.

To ensure reliable results, always fire from the prone position.

Look through the scope, concentrating on the center aiming point of the reticle. Move head slightly up and down. The aiming point should remain in exactly the same position against the target. If it moves, turn the focus adjustment dial until it becomes stable.

Windage And Elevation Adjustment

1. Set the scope to the highest power. Acquire the target. Fire one round.

2. Make appropriate adjustments to move center point of aim.

NOTE

Use center of bullet strike as a reference point for the final adjustments to windage and elevation. One click is equal to 1/4 minute of angle.

3. Fire a second round using the same point of aim, confirm bullet strike is the same as the point of aim. If not, continue adjustment until zeroed.

NOTE

All dials on the scope either numbered or with an indicator can be repositioned to align the marked zero of the dial with the position indicator without changing the adjustment setting of the scope that was achieved when zeroing. This allows the shooter to know the original zero of the rifle in the event that further adjustments are required in the field.

4. To reposition the dials, remove the turret caps.

DAYLIGHT SCOPE - Continued

5. Loosen the setscrews that surround the top of the knob until the cylinder turns freely. Perform this step for elevation and windage adjustment knobs.

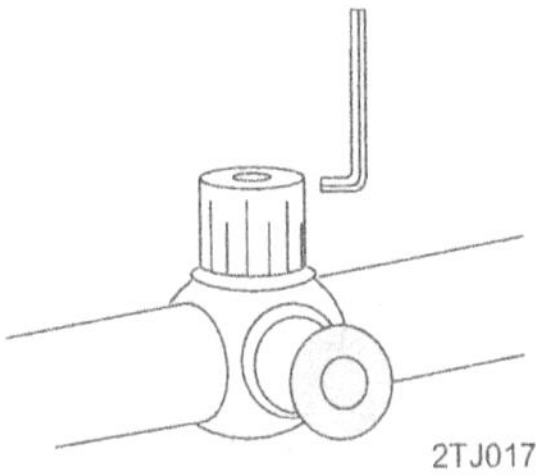

6. Loosen allen screws, zero line on the cylinder and zero line on the cap must be aligned.

7. The power selector ring is located in front of the eyepiece assembly. Turn the ring to align the number indicating the desired magnification with the gold dot on the body of the scope. Tighten setscrews.

Use Of Reticle To Estimate Range

The reticle for the M107 telescope can be used to estimate range when the size of the target is known. The following dimensions are related to Minutes of Angle (MOA) and can be used in range estimation. One mil = 3.438 MOA.

Dimension	MOA
A to F	1.0
B	0.1
C	0.7
D	3.6
E	18.0

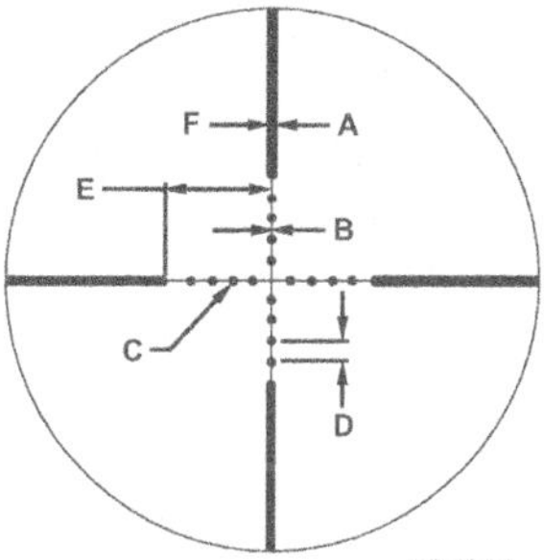

As a point of reference, the Mil-Dot "C" will cover a quarter at 100 meters. This means that at 200 meters, the quarter will be the size of 1/2 of the dot "C", at maximum power.

DAYLIGHT SCOPE - Continued

Level Adjustment

Each time the rifle is fired, the vertical crosshair should be held in the vertical position. Canting the rifle to the left will cause the rifle to shoot to the right. Conversely it will shoot left when the rifle is canted to the right. Obtain the correct shooting position.

Position (1) indicates the rifle is canted too far to the right, position (2) shows the rifle is canted too far to the left, and position (3) is the correct sight picture for the rifle held vertically.

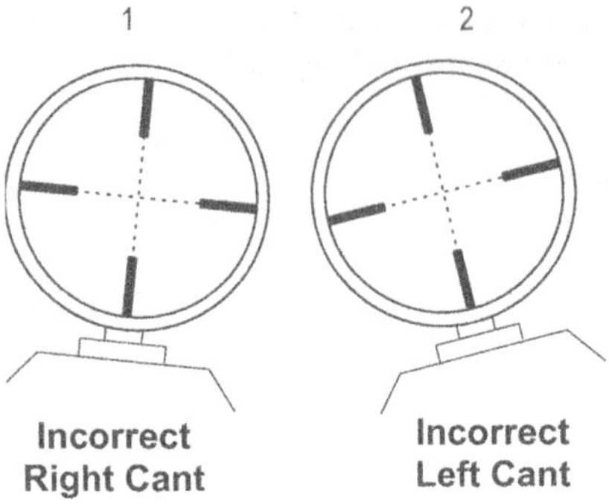

END OF WORK PACKAGE

OPERATOR
LONG RANGE SNIPER RIFLE, M107
NSN 1005-01-469-2133
OPERATION UNDER UNUSUAL CONDITIONS

INITIAL SETUP:

References

WP 0012 00

EMERGENCY PROCEDURES - IMMEDIATE ACTION

An emergency condition exists when the magazine for the rifle has been lost or damaged and is not available.

In an emergency, the weapon may be fired without the magazine. A single cartridge may be inserted directly into the chamber. Insert round through magazine well into chamber. After insertion, close and lock the bolt to fire.

NOTE

Unusual conditions are defined as any condition requiring special maintenance of the rifle. Perform the maintenance outlined for the climate that most applies to your operational area. Refer to WP 0012 00 for lubrication instructions.

PERFORM SPORTS

Immediate action is the prompt action taken by the user to correct a stoppage. The procedure for applying immediate action should become instinctive to the user, without the user attempting to discover the cause. It is important that the user apply immediate action instinctively to correct a stoppage.

Slap the magazine
Pull charging handle
Observe clear chamber
Release bolt forward
Tap seat the bolt
Squeeze the trigger

CAUTION

If extensive corrosion is found and cleaning does not solve the problem, turn the complete weapon system in to the proper maintenance/supply channel.

END OF WORK PACKAGE

CHAPTER 3

TROUBLESHOOTING PROCEDURES FOR LONG RANGE SNIPER RIFLE, M107

OPERATOR
LONG RANGE SNIPER RIFLE, M107
NSN 1005-01-469-2133
TROUBLESHOOTING INDEX

GENERAL

The malfunction/symptom index can be used to as a quick guide to troubleshooting. Common malfunctions are listed in cycle of function order with a work package page reference to the troubleshooting table where a test or inspection and corrective action are provided.

This manual cannot list all malfunctions that may occur, nor all tests or inspections and corrective actions. If a malfunction is not listed or is not corrected by listed corrective actions, notify organizational maintenance.

MALFUNCTION/SYMPTOM INDEX

END OF WORK PACKAGE

OPERATOR

LONG RANGE SNIPER RIFLE, M107
NSN 1005-01-469-2133
TROUBLESHOOTING PROCEDURES

INITIAL SETUP:

References
WP 0005 00
WP 0014 00

GENERAL

1. Check the mount. Make sure the scope is mounted securely to the rifle. Try, with bare hands, to twist the scope in the rings. If there is movement, retighten the mounting system.

2. Ensure that the weapon is properly assembled and the upper and lower receiver assemblies have positive retention of the midlock and rear lock pins.

GENERAL - Continued

3. Be sure to use proper ammunition, of the same bullet type, weight, and preferably lot number. Use of different types of ammunition may have different results and cause inaccuracy in adjustments.

4. Be sure that both the barrel and chamber are clean. Heavy grease on the rifle and copper fouling can diminish the accuracy of the rifle.

The table lists the common malfunctions, which you may find during the operation or maintenance of the rifle, or its components. You should perform the tests/inspections and corrective actions in the order listed.

LONG RANGE SNIPER RIFLE, M107

Table 1. Troubleshooting Procedures.

MALFUNCTION	TEST OR INSPECTION	CORRECTIVE ACTION
1. Failure to feed.	1. Verify sluggish action.	1. Clean and lubricate or (if cold) check over lubrication.
	2. Check for improper seating of magazine.	1. Reinsert magazine properly (see WP 0005 00).
2. Failure to chamber.	1. Check for damaged cartridge.	1. Remove cartridge and recharge (reload) (see WP 0005 00).
	2. Check for dirty chamber.	2. Clear and clean chamber.

LONG RANGE SNIPER RIFLE, M107 - Continued

Table 1. Troubleshooting Procedures - Continued.

MALFUNCTION	TEST OR INSPECTION	CORRECTIVE ACTION
3. Failure to lock or unlock	1. Check for obstruction between firing pin and bolt.	1. Disassemble and clean firing pin and bolt.
		2. Clean firing pin and bolt, try different ammo.
	2. Check for blown primer wedged between firing pin and bolt.	1. Evacuate to direct support maintenance.
4. Failure to fire.	1. Verify faulty ammunition.	1. Replace ammunition.

Table 1. Troubleshooting Procedures - Continued.

MALFUNCTION	TEST OR INSPECTION	CORRECTIVE ACTION
4. Failure to fire - Continued	2. Check for improper installation of firing mechanism	2. Assemble firing mechanism properly.
	3. Check for broken firing pin.	3. Turn-in firing pin for repair.
	4. Check for broken or weakened firing spring(s).	4. Turn-in firing spring for repair.
	5. Check for bent or broken cocking piece.	5. Turn-in cocking piece for repair.

LONG RANGE SNIPER RIFLE, M107 - Continued

Table 1. Troubleshooting Procedures - Continued.

MALFUNCTION	TEST OR INSPECTION	CORRECTIVE ACTION
5. Failure to extract.	1. Check for broken extractor.	1. Replace extractor (see WP 0014 00).
	2. Verify extractor is not moving freely in slot.	2. Remove extractor and clean.
	3. Check for dirty chamber.	3. Clean chamber.
6. Failure to eject.	1. Check for frozen or damaged ejector or spring.	1. Turn-in ejector or spring for repair.

Table 1. Troubleshooting Procedures - Continued.

MALFUNCTION	TEST OR INSPECTION	CORRECTIVE ACTION
7. Very hard recoil.	1. Check for faulty/hot ammunition.	1. Replace or cool ammunition.
	2. Check for missing or damaged/clogged muzzle brake.	2. Turn-in muzzle brake for repair.
8. Failure to cock		1. Evacuate to Direct Support Maintenance.

END OF WORK PACKAGE

CHAPTER 4

MAINTENANCE INSTRUCTIONS
FOR
LONG RANGE SNIPER RIFLE, M107

OPERATOR
LONG RANGE SNIPER RIFLE, M107
NSN 1005-01-469-2133
PREVENTIVE MAINTENANCE CHECKS AND SERVICES (PMCS)
INTRODUCTION

GENERAL

Always observe the **WARNINGS** and **CAUTIONS** appearing in your PMCS table. **WARNINGS** and **CAUTIONS** appear before applicable procedures. You must observe these **WARNINGS** and **CAUTIONS** to prevent serious injury to yourself and others or to prevent your equipment from being damaged.

EXPLANATION OF COLUMN ENTRIES

ITEM NUMBER column. Numbers in this column are for reference. When completing DA Form 2404 (Equipment Inspection and Maintenance Worksheet), include the item number for the check/service indicating a fault. Item numbers also appear in the order that you must do checks and services for the intervals listed.

INTERVAL column. This column tells you when you must do the procedure in the procedure column. BEFORE procedures must be done before you operate or use the equipment for its intended mission. DURING procedures must be done during the time you are operating or using the equipment for its intended mission. AFTER procedures must be done immediately after you have operated or used the equipment.

ITEM TO BE CHECKED OR SERVICED column. This column lists the items to be checked or serviced.

PROCEDURE column. This column gives the procedure you must do to check or service the item listed in the ITEM TO BE CHECKED OR SERVICED column to know if the equipment is ready or available for its intended mission or for operation. You must do the procedure at the time stated in the interval column.

NOT FULLY MISSION CAPABLE IF: column. Information in this column tells you what faults will keep your equipment from being capable of performing its primary mission. If you do check/service procedures that show faults listed in this column, do not operate the equipment. Follow standard operating procedures for maintaining the equipment or reporting equipment failure.

END OF WORK PACKAGE

OPERATOR
LONG RANGE SNIPER RIFLE, M107
NSN 1005-01-469-2133
PREVENTIVE MAINTENANCE CHECKS AND SERVICES (PMCS) INCLUDING LUBRICATION INSTRUCTIONS

INITIAL SETUP:

Tools and Special Tools

Cleaning Rod (item 47, WP 0021 00)

Materials/Parts

Cleaner, Lubricant and Preservative (CLP) (item 8, WP 0022 00)
Lubricating Oil (LAW) (item 13, WP 0022 00)
Lubricating Oil (LSA) (item 14, WP 0022 00)

Materials/Parts - Continued

Lubricating Oil (LSAT) (item 15, WP 0022 00)
Small Arms Cleaning Swab (item 18, WP 0022 00)
Wiping Rag, (item 17, WP 0022 00)

References

WP 0018 00

PMCS PROCEDURES

Table 1. Preventive Maintenance Checks and Services for Long Range Sniper Rifle, M107.

ITEM NO.	INTERVAL	ITEM TO BE CHECKED OR SERVICED	PROCEDURE	EQUIPMENT NOT READY/ AVAILABLE IF:
1	Before	M107	Hand function the weapon to ensure it is functional and visually check the exterior of the weapon and components for damage. Check components for cracks, breaks, and damage. If any faults are found, notify unit maintenance.	Parts missing, damaged, or broken.

Table 1. Preventive Maintenance Checks and Services for Long Range Sniper Rifle, M107 - Continued.

ITEM NO.	INTERVAL	ITEM TO BE CHECKED OR SERVICED	PROCEDURE	EQUIPMENT NOT READY/ AVAILABLE IF:
2	Before	Muzzle Brake	Check to see that both retaining screws are secure and that muzzle brake is level to the ground.	Muzzle brake is loose or crooked.
3	Before	Barrel Assembly	Check to ensure bore is free of obstruction. Check for excess lubrication in bore area. Swab dry.	Bore is obstructed.
4	Before	Scope Mounting Hardware	Check to see that all hardware is tight and that scope is secure to weapon.	Scope is loose or hardware is missing.

PMCS PROCEDURES – Continued

Table 1. Preventive Maintenance Checks and Services for Long Range Sniper Rifle, M107 - Continued.

ITEM NO.	INTERVAL	ITEM TO BE CHECKED OR SERVICED	PROCEDURE	EQUIPMENT NOT READY/ AVAILABLE IF:
5	Before	Lock pins	Check to see that the rear and midlock pins are installed so that the retaining bearing is visible on the opposite side of the receiver.	Pin cannot be inserted far enough for bearing to be exposed. Pin missing.
6	Before	Magazine	Ensure that the magazine has free travel of the follower and that the magazine tube is not damaged (bent or cracked).	Free travel of follower is not present or magazine is damaged.

Table 1. Preventive Maintenance Checks and Services for Long Range Sniper Rifle, M107 - Continued.

ITEM NO.	INTERVAL	ITEM TO BE CHECKED OR SERVICED	PROCEDURE	EQUIPMENT NOT READY/ AVAILABLE IF:
6.1	Before and After	Magazine	Ensure magazine is clean	Dirty, or has lubricant in it.
7	During	Muzzle Brake	Check to see that both retaining screws are secure and that muzzle brake is level to the ground.	Muzzle brake is loose or crooked.
8	During	Scope Mounting Hardware	Check to see that all hardware is tight and that scope is secure to weapon.	Scope is loose or hardware is missing.

PMCS PROCEDURES – Continued

Table 1. Preventive Maintenance Checks and Services for Long Range Sniper Rifle, M107 - Continued.

ITEM NO.	INTERVAL	ITEM TO BE CHECKED OR SERVICED	PROCEDURE	EQUIPMENT NOT READY/ AVAILABLE IF:
9	During	Lock pins	Check to see that the rear and midlock pins are installed so that the retaining bearing is visible on the opposite side of the receiver.	Pin cannot be inserted far enough for bearing to be exposed. Pin missing.
10	During	Magazine	Ensure that the magazine has free travel of the follower and that the magazine tube is not damaged (bent or cracked).	Free travel of follower is not present or magazine is damaged.

Table 1. Preventive Maintenance Checks and Services for Long Range Sniper Rifle, M107 - Continued.

ITEM NO.	INTERVAL	ITEM TO BE CHECKED OR SERVICED	PROCEDURE	EQUIPMENT NOT READY/ AVAILABLE IF:
11	After	M107	Hand function the weapon to ensure it is functional and visually check the exterior of the weapon and components for damage. Check components for cracks, breaks, and damage. If any faults are found, notify unit maintenance.	Parts missing, damaged, or broken.

PMCS PROCEDURES – Continued

Table 1. Preventive Maintenance Checks and Services for Long Range Sniper Rifle, M107 - Continued.

ITEM NO.	INTERVAL	ITEM TO BE CHECKED OR SERVICED	PROCEDURE	EQUIPMENT NOT READY/ AVAILABLE IF:
12	After	Muzzle Brake	Check to see that both retaining screws are secure and that muzzle brake is level to the ground.	Muzzle brake is loose or crooked.
13	After	Scope Mounting Hardware	Check to see that all hardware is tight and that scope is secure to weapon.	Scope is loose or hardware is missing.
14	After	Clean Barrel Bore	Clean chamber and barrel bore immediately after firing.	Chamber and/or barrel obstructed.

Table 1. Preventive Maintenance Checks and Services for Long Range Sniper Rifle, M107 - Continued.

ITEM NO.	INTERVAL	ITEM TO BE CHECKED OR SERVICED	PROCEDURE	EQUIPMENT NOT READY/ AVAILABLE IF:
15	After	Lock pins	Check to see that the rear and midlock pins are installed so that the retaining bearing is visible on the opposite side of the receiver.	Pin cannot be inserted far enough for bearing to be exposed. Pin missing.
16	After	Magazine	Ensure that the magazine has free travel of the follower and that the magazine tube is not damaged (bent or cracked).	Free travel of follower is not present or magazine is damaged.

LUBRICATION INSTRUCTIONS

WARNING

Never mix lubricants on the weapon, always completely remove one lubricant before using another. Mixing lubricants can cause viscosity change, resulting in weapon damage or malfunction, which can cause injury.

1. LSAT (lubricant, small arms (with Teflon)), LSA, and CLP (cleaner, lubricant, and preservative) are the authorized lubricants to use on your rifle at normal temperatures to -10°F (-27°C). At temperatures below -10°F (-27°C) use LAW.

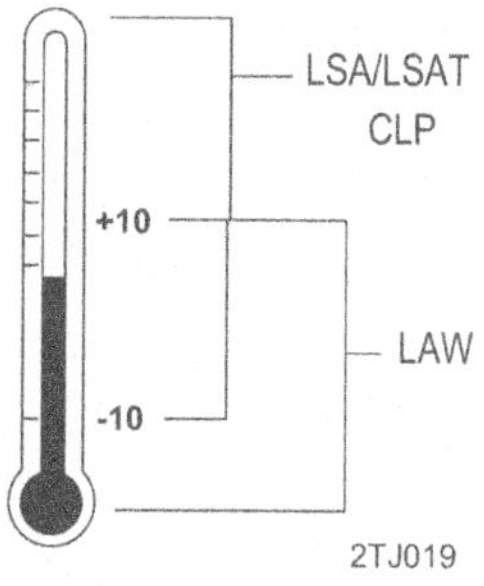

NOTE

Remember to remove excessive oil from the bore before firing.

2. Lightly Lube - A film of oil barely visible to the eye.
3. Generously Lube - Heavy enough so that it can be spread with finger.

CAUTION

When operating rifle in extremely cold climates, clean and lubricate the rifle inside at room temperature, if possible using LAW.

Climates

1. Hot, Dusty, and Sandy Areas.
   a. Clean often. Lightly lube.
   b. Wipe oil from exposed surfaces with clean wiping rag (item 17, WP 0022 00).
   c. Keep sand out of parts.

LUBRICATION INSTRUCTIONS - Continued

2. Extremely Cold Climate.
   a. Use LAW (item 13, WP 0022 00).
   b. Keep dry.
   c. Use LAW (item 13, WP 0022 00) lightly.
3. Hot, Wet Climate.
   a. Use LSAT, LSA, or CLP (item 15, 14, or 8, WP 0022 00) and inspect often.
   b. Use LSAT, LSA, or CLP (item 15, 14, or 8, WP 0022 00) lightly.
   c. Keep rifle dry.

External Surfaces

Put LSAT, LSA, or CLP (item 15, 14, or 8, WP 0022 00) on a clean swab (item 18, WP 0022 00) and generously lubricate:

- Bolt – locking lugs and cam slot;
- Bolt Carrier – receiver bearing surfaces;
- Barrel – bolt locking surfaces,
- Receiver – rails that bolt carrier rides on.

All Other Areas

Lightly lubricate - including bore.

NOTE

LSAT, a multi-purpose lubricant containing Teflon, has displayed exceptional lubrication performance as well as resistant to collecting sand.

Medium-Heavy Lubrication

Generously lubricate (meaning obvious to visual inspection) the following areas:

- Behind the bolt locking lugs.
- Bolt body.
- Bolt cam pin.
- Moving parts of the bolt and carrier assemblies.
- The receiver rails which the bolt and carrier assemblies ride in, in the lower receiver assembly.
- The barrel extension where it rides in the weapon.

LUBRICATION INSTRUCTIONS - Continued

Before Firing

1. Using a clean wiping rag (item 17, WP 0022 00), wipe bore dry.

2. Ensure the weapon is properly lubricated.

3. After exposure to water, make sure the rifle is dry before lubricating. Disassemble, clean, lightly oil and assemble as soon as possible.

4. In extreme cold, perform the following steps:

a. When operating rifle in extremely cold climate clean and lubricate rifle inside at room temperature if possible.

b. Apply a light coat of Lubricant Arctic Weather (LAW) (item 13, WP 0022 00) to all functional parts.

c. To prevent freezing, keep the rifle covered when moving from a warm to a cold area. This will allow gradual cooling.

d. Always keep the rifle dry.

e. Keep ammunition dry; moisture will cause malfunctions. Do not lubricate ammunition.

f. Always keep snow out of the bore of the barrel. If snow should get into the bore, clean the bore before firing, using a swab (item 18, WP 0022 00) and cleaning rod (item 47, WP 0021 00).

5. In hot, dry climates, perform the following steps:

NOTE

Dust and sand can get into rifle and cause malfunctions and excessive wear on component contact surfaces during firing.

a. Keep the rifle covered when possible.

b. Use CLP, LSA, or LSAT (item 8, 14, or 15, WP 0022 00) sparingly.

6. For heavy rain and fording operations – all climates, perform the following steps:

a. Perform maintenance in accordance with cleaning procedures in WP 0018 00 and apply a thin coat of CLP, LSA, LSAT (item 8, 14, or 15, WP 0022 00). Do NOT lubricate ammunition.

LUBRICATION INSTRUCTIONS - Continued

b. Always attempt to keep rifle dry.

c. Always drain any water from barrel prior to firing. Dry the bore with a swab (item 18, WP 0022 00) and cleaning rod (item 47, WP 0021 00).

7. For hot, wet climates, perform the following steps:

a. Perform maintenance more frequently. Inspect hidden surfaces for corrosion. If corrosion is found, clean and lubricate with CLP, LSA, or LSAT (item 8, 14, or 15, WP 0022 00).

b. To help prevent corrosion, remove handprints with wiping rag (item 17, WP 0022 00). Dry off and then lubricate the rifle.

END OF WORK PACKAGE

OPERATOR
LONG RANGE SNIPER RIFLE, M107
NSN 1005-01-469-2133
INSPECTION OF INSTALLED ITEMS

INTRODUCTION

An inspection shall be performed on components, assemblies, or parts installed on the M107 Sniper Rifle. The purpose of this inspection is to determine if the item is damaged or deteriorated to the extent that it should be replaced or repaired. Table 1 lists the required inspections and acceptable/unacceptable conditions for the M107 Sniper Rifle.

INSPECTION OF INSTALLED ITEMS

Table 1. Inspection of Installed Items on M107 Sniper Rifle.

ASSEMBLY	ITEM INSPECTED	PROCEDURE	CONDITION
Upper Receiver Assembly	Upper Receiver Assembly	Visual inspection (special attention should be paid to the hinge lip at the front of the receiver).	Upper receiver assembly should NOT be cracked, bent or burred.
	Barrel Springs	Visual inspection.	Barrel springs must NOT be over-stretched, and each coil should be tight, with NO spaces between coils when barrel springs are relaxed.
	Barrel	Visual inspection.	Barrel should be clean and free of obstruction.

Table 1. Inspection of Installed Items on M107 Sniper Rifle - Continued.

ASSEMBLY	ITEM INSPECTED	PROCEDURE	CONDITION
Upper Receiver Assembly - Continued	Impact Bumpers	Visual inspection.	Impact bumpers should be in good condition, not frayed, cracked, or twisted.
	Front Sight	Traverse through complete range of motion	Should have fluid motion throughout traverse and have positive retention in the upright position.
Barrel Assembly	Muzzle Brake	Visual inspection and check for looseness.	Muzzle brake should be tight and fully screwed on, and properly positioned.
	Scope Mountings	Visual inspection and check for looseness	Scope mountings should be tight and in good condition.

INSPECTION OF INSTALLED ITEMS - Continued

Table 1. Inspection of Installed Items on M107 Sniper Rifle - Continued.

ASSEMBLY	ITEM INSPECTED	PROCEDURE	CONDITION
Barrel Assembly - Continued	Iron Sights	Traverse iron sight through complete range and lightly lubricate iron sight to prevent corrosion.	Iron sights should traverse through complete range without binding or stoppage.
Bolt Assembly	Extractor and Ejector	Check spring tension and visually inspect for chips or wear.	Extractor and ejector springs must NOT be over-stretched. There should be no chips and wear on the extractor or ejector.

Table 1. Inspection of Installed Items on M107 Sniper Rifle - Continued.

ASSEMBLY	ITEM INSPECTED	PROCEDURE	CONDITION
Bolt Assembly - Continued	Firing Pin	Push the bolt into the carrier, using the midlock pin, de-cock before inspecting firing pin protrusion. Check firing pin hole (on bolt face) for erosion/pitting.	Firing pin hole should not have erosion/pitting.
	Bolt Latch	Visual inspection.	Bolt latch should not show any deformation.
	Sear	Swing the cocking lever forward.	The sear should capture the firing pin extension before the cocking lever is fully depressed.

INSPECTION OF INSTALLED ITEMS - Continued

Table 1. Inspection of Installed Items on M107 Sniper Rifle - Continued.

ASSEMBLY	ITEM INSPECTED	PROCEDURE	CONDITION
Carrier Assembly	Bolt Assembly	With firing mechanism de-cocked, depress bolt latch and manually work the bolt in and out, feeling for any roughness, which may indicate wear, corrosion or grit in the carrier.	Bolt assembly should be smooth.
Lower Receiver Assembly	Mainspring	With the bolt and carrier assemblies in position, pull them rearward and check to see that the mainspring moves freely (full travel).	If there is ANY binding, take the weapon to the armorer for further inspection.

Table 1. Inspection of Installed Items on M107 Sniper Rifle - Continued.

ASSEMBLY	ITEM INSPECTED	PROCEDURE	CONDITION
Lower Receiver Assembly - Continued	Mainspring Housing	Hold the bolt and carrier assemblies back and down approximately 1/4 in. inside the receiver. With the thumb safety on 'FIRE', pull the trigger.	Firing mechanism should function. If the housing is bent, the bolt and carrier assemblies will rise as the trigger is pulled, preventing proper function.
	Lower Receiver Assembly	Visual inspection.	Lower receiver assembly should NOT be cracked, bent or burred.
	Bipod Assembly	Check bipod assembly and mounting hardware.	Legs should extend and hardware functions properly.

END OF WORK PACKAGE

OPERATOR
LONG RANGE SNIPER RIFLE, M107
NSN 1005-01-469-2133
FIELD STRIPPING

INTRODUCTION

The M107 is field stripped into four major components: upper receiver assembly (1), lower receiver assembly (2), bolt and carrier assemblies (3), and the telescopic sight assembly (4).

NOTE

Unless there is something wrong with the telescope, it will not be removed for normal field stripping.

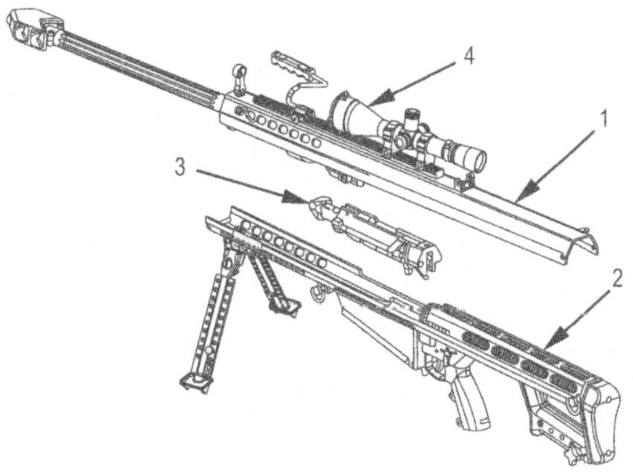

FIELD STRIPPING

WARNING

Ensure that the weapon is unloaded and on SAFE before performing these procedures.

1. Remove rear lock (1) and midlock (2) pins.

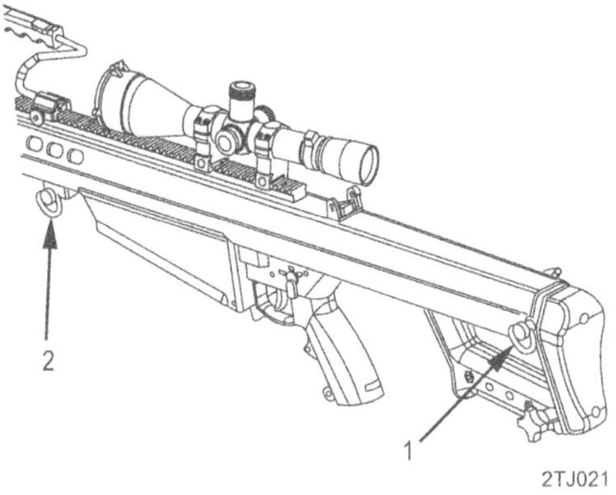

2. Grasp charging handle (3) on the bolt carrier (4) and pull to the rear until the bolt (5) clears the barrel extension.

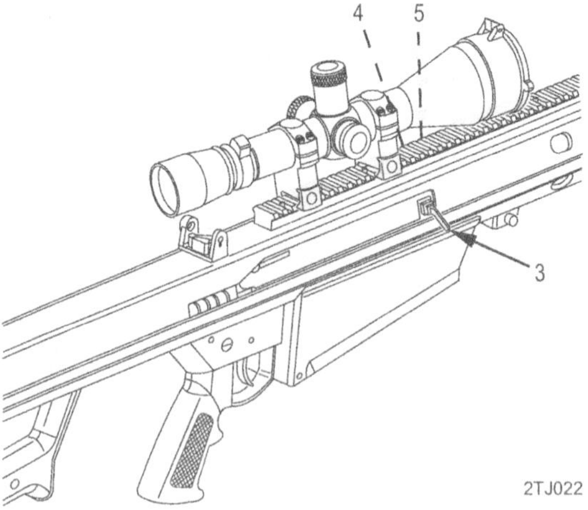

FIELD STRIPPING - Continued

3. While holding the charging handle (3) to the rear, lift the back end of the upper receiver assembly (6) until it clears the bolt (5). Slowly allow the bolt carrier (4) to return to its forward position.

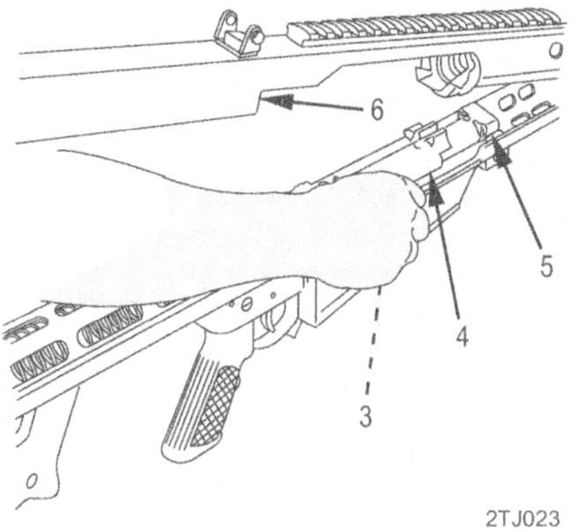

4. Disengage front hook (7) from the front hook pin (8) on the lower receiver (9) and lift upper receiver assembly (6) clear of lower receiver assembly.

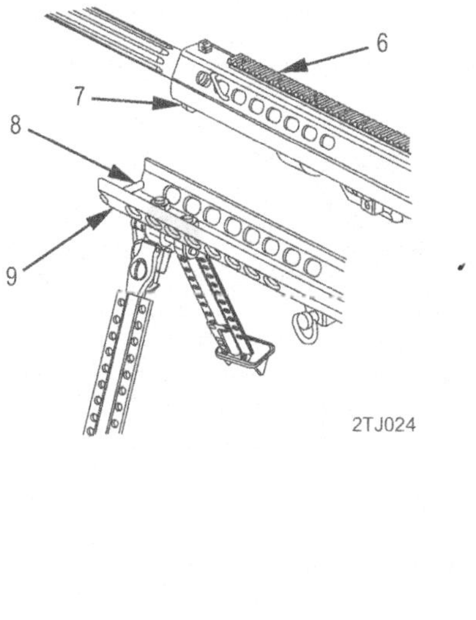

FIELD STRIPPING - Continued

WARNING

Buffer and buffer spring are under heavy spring tension.

6. Pull the bolt carrier (4) rearward and insert the rear lock pin (1) through the buffer (10) and buffer spring (11).

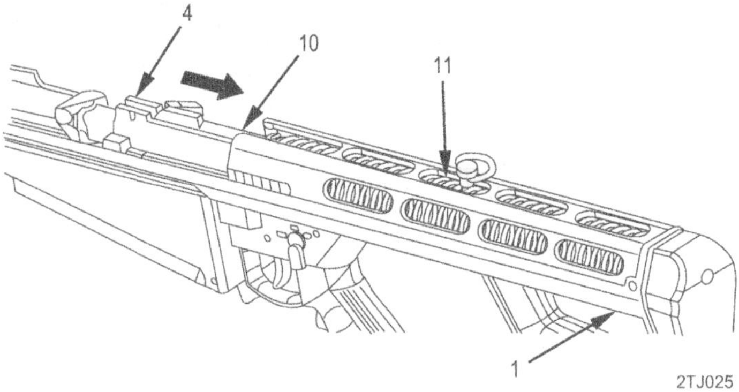

2TJ025

CAUTION

When removing the bolt carrier from the lower receiver assembly, ensure carrier is completely forward of the housing before lifting bolt to avoid damaging lower receiver assembly.

7. Gently pull the bolt carrier (4) forward and lift out of the lower receiver assembly (9).

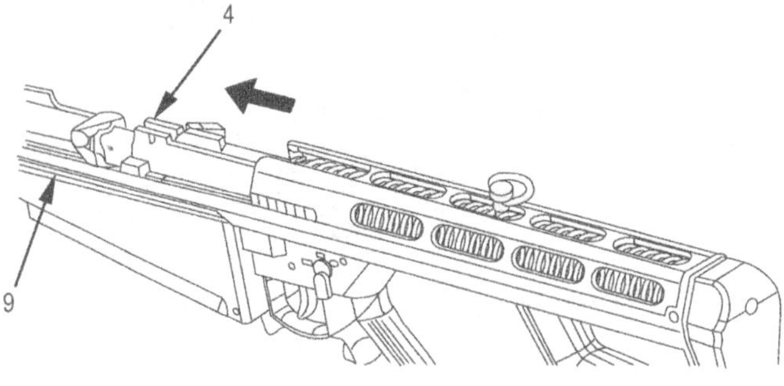

2TJ026

END OF WORK PACKAGE

OPERATOR
LONG RANGE SNIPER RIFLE, M107
NSN 1005-01-469-2133
EXTRACTOR MAINTENANCE

INITIAL SETUP:

Tools and Special Tools

Deployment Tool Kit (item 6, WP 0021 00)

REMOVAL

WARNING

Always point the bolt away from face and eyes to avoid possible injury if parts fly free.

CAUTION

Cover plunger and plunger spring with thumb while sliding extractor out of slot to prevent loss or damage.

NOTE

Perform Removal procedure only upon extractor, extractor plunger or extractor spring failure.

1. To remove the extractor (1) from the bolt (2), depress the extractor plunger (3) by inserting a 1/16 in. punch through the extractor hole (4).

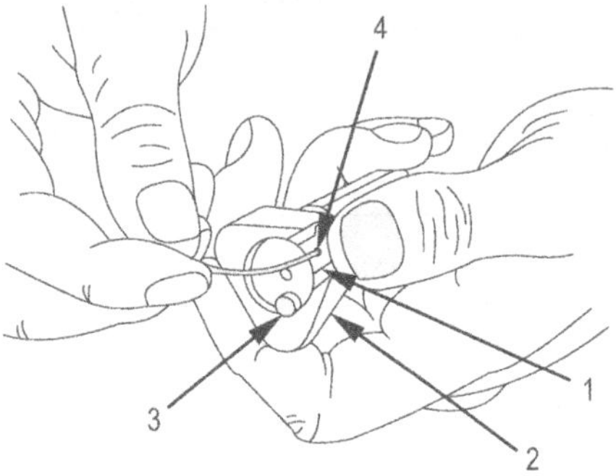

2TJ027

REMOVAL - Continued

2. Slide the extractor (1) out of the slot. Remove extractor plunger (3) and extractor spring (5). Visually inspect all components for damage.

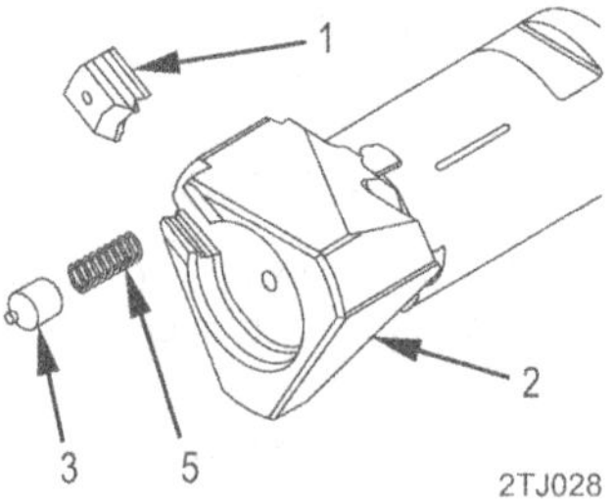

2TJ028

INSTALLATION

1. Insert extractor spring (5) and extractor plunger (3) into bolt (2).

2. Slide the extractor (1) into the slot.

END OF WORK PACKAGE

OPERATOR

LONG RANGE SNIPER RIFLE, M107

NSN 1005-01-469-2133

MAINSPRING AND MAINSPRING BUFFER MAINTENANCE

REMOVAL

1. Using your finger, push the mainspring buffer (1) to the rear and place the rear lock pin (2) through the coils of the mainspring (3) and the ports in the mainspring housing (4).

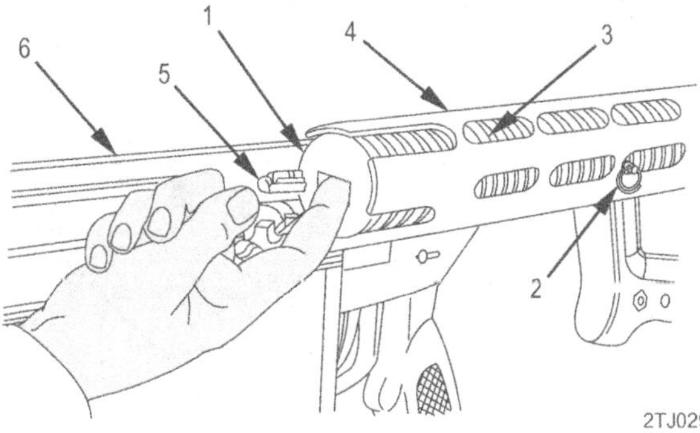

2. Place your finger into the slot on the lower end of mainspring buffer (1), and turn the mainspring buffer so that the groove on the flange lines up with the buffer stop (5) on the lower receiver assembly (6). Remove the rear lock pin (2). Slowly and carefully remove the mainspring buffer (1) and mainspring (3).

INSTALLATION

1. Slide the mainspring (1) into the lower receiver assembly (2). Place the mainspring buffer (3) into the mainspring. Guide the mainspring into the housing (4) until the mainspring buffer is even with the housing. Place your finger into the slot on the mainspring buffer, and turn the mainspring buffer so that the grove in its flange lines up with the buffer stop (5) on the lower receiver assembly.

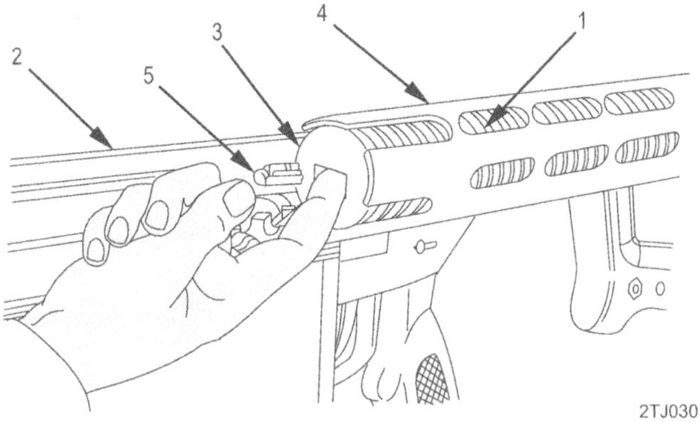

2TJ030

INSTALLATION - Continued

2. Push the mainspring buffer (3) to the rear of the housing (4) and secure it by placing the rear lock pin (6) through coils of the mainspring (1) and the ports in the mainspring housing. Using your finger, ensure that the groove and slot are NOT aligned. Remove the rear lock pin. Ease off the pressure until the mainspring buffer stops on the buffer stop (5).

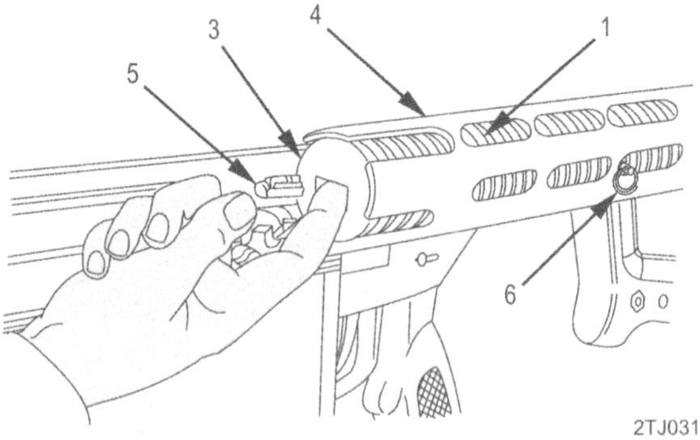

2TJ031

END OF WORK PACKAGE

OPERATOR
LONG RANGE SNIPER RIFLE, M107
NSN 1005-01-469-2133
BARREL MAINTENANCE

ASSEMBLY

1. Carefully pick up the upper receiver assembly (1). The barrel (2) will be nested inside the receiver for compact storage and transit. Move the impact bumpers (3) into position on either side of the receiver's central barrel bushing, so that they rest snugly against the bushing.

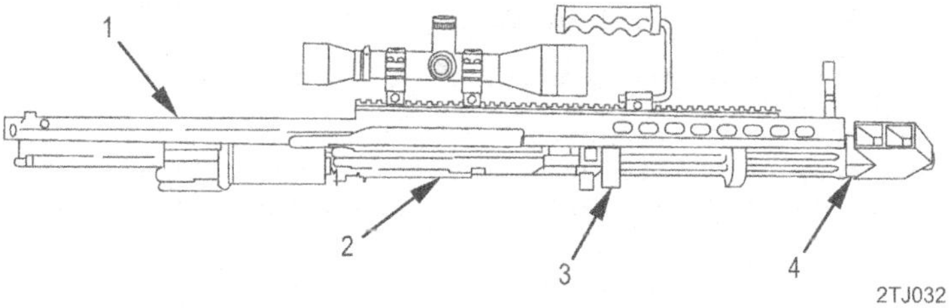

2. Align the barrel (2) so that its feed ramp (slanted entry to the firing chamber) is on the bottom. Keeping fingers away from the barrel, hold the upper receiver assembly (1) horizontally, then tilt it in the direction of the muzzle (4). The barrel should fall into place, at its full forward extension into the upper receiver assembly.

WARNING

The tension on the barrel springs is about 70 pounds. Serious injury may result if springs are released suddenly.

Incomplete or improper assembly may result in injury.

3. The barrel springs (5) at the front of the upper receiver assembly (1) are held together by the barrel key (6) which acts as a spring yoke. Maintaining the downward tilt of the upper receiver assembly (to keep the barrel in place) firmly grasp the barrel key-not the springs-and pull it into place on the forward slot of the barrel. Work the barrel key from side to side until it is firmly seated in the barrel slot. The upper receiver assembly is now fully assembled.

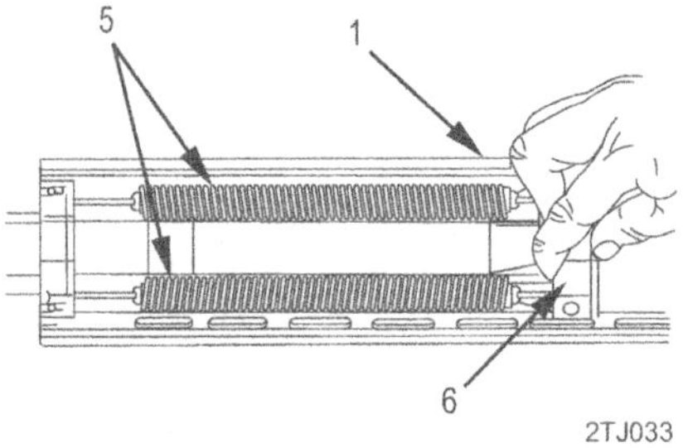

2TJ033

STORAGE

CAUTION

Do not pull on barrel springs to remove the barrel key. Doing so may damage the springs.

Barrel spring is under tension.

1. To nest the barrel (1) for storage and transit, it is necessary to remove the barrel key (2) from the barrel key slot.

2. Withdraw the barrel key (2) from the slot in the barrel (1) by slowly working it out. Slide the barrel out the rear of the upper receiver assembly (3).

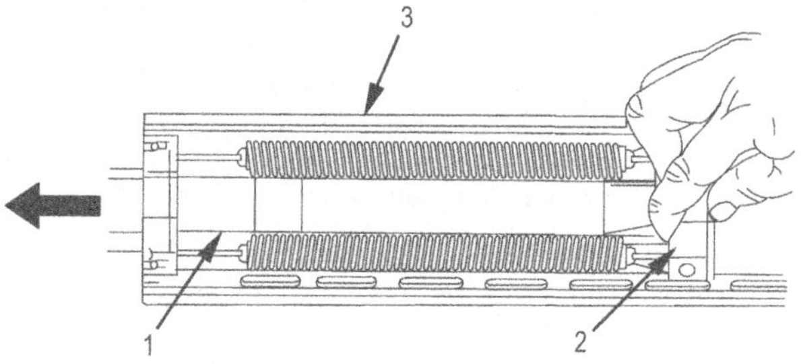

END OF WORK PACKAGE

OPERATOR

LONG RANGE SNIPER RIFLE, M107

NSN 1005-01-469-2133

BOLT AND CARRIER ASSEMBLIES MAINTENANCE

INITIAL SETUP:

Tools and Special Tools

Deployment Tool Kit (item 6, WP 0021 00)

DISASSEMBLY

WARNING

If the spring-loaded cam is lifted too far, the spring may lose tension. If this occurs, the weapon could malfunction, or it could allow the weapon to fire when unlocked, with the potential for serious injury.

1. If the bolt assembly (1) and bolt spring (2) are being removed separately, begin by using the rear lock pin or a 1/8 inch punch to disengage the cam pin assembly (3).

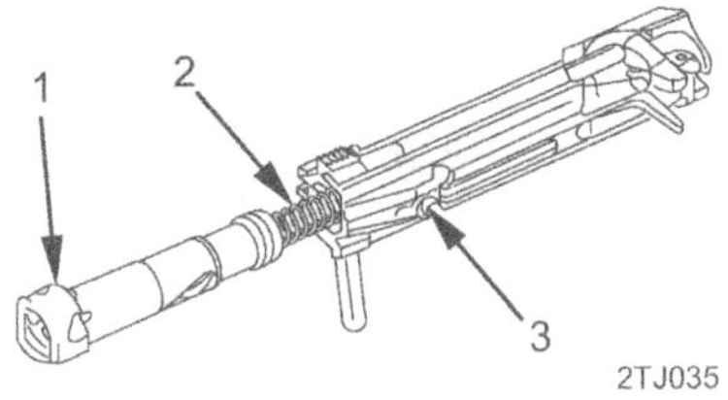

2. Lift or pry the cam pin spring (4) up just far enough to clear the cam groove. Depress the bolt latch (5) on its rearward portion. At this point the bolt assembly (1) should spring forward. Grasp the bolt assembly and remove it from the carrier (6), being careful not to lose or deform the bolt spring (2).

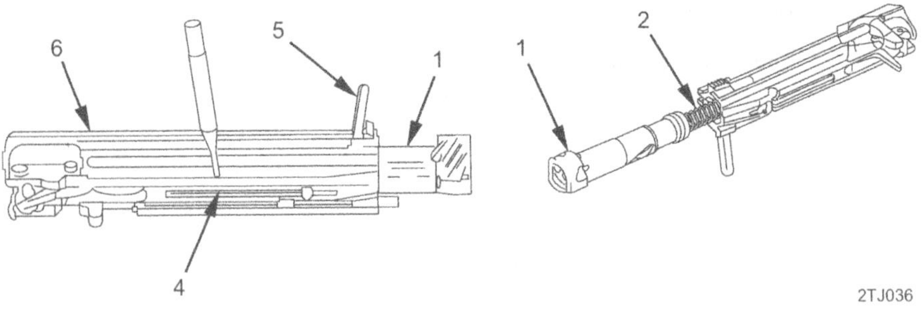

ASSEMBLY

While lifting cam pin assembly (4), insert the bolt assembly (1) into position. The bolt assembly must be compressed against the bolt spring (2) until the cam (3) slips into the cam groove. Release cam pin assembly (4).

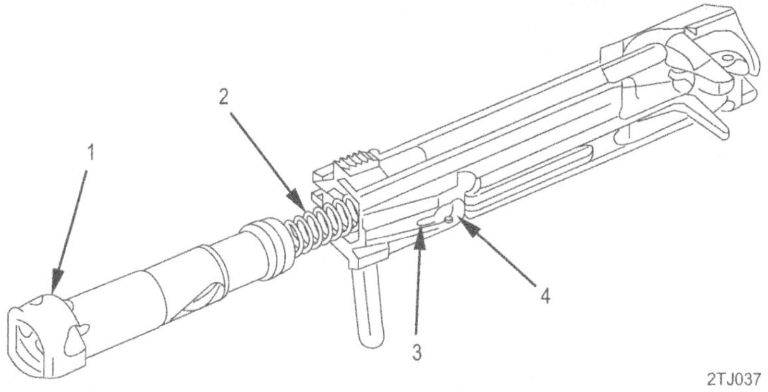

END OF WORK PACKAGE

OPERATOR

LONG RANGE SNIPER RIFLE, M107

NSN 1005-01-469-2133

CLEANING INSTRUCTIONS

INITIAL SETUP:

Tools and Special Tools

Deployment Cleaning Kit (item 8, WP 0021 00)

Deployment Parts Kit (item 7, WP 0021 00)

Deployment Tool Kit (item 6, WP 0021 00)

Tools and Special Tools - Continued

Five Piece Cleaning Rod (item 10, WP 0021 00)

Optics Cleaning Kit (item 9, WP 0021 00)

DEPLOYMENT KIT

The Deployment kit is made up of five separate kits, they are:

- Weapons cleaning kit
- Optics cleaning kit
- Deployment tool kit
- Deployment parts kit
- Five piece cleaning rod

DEPLOYMENT KIT – Continued

Weapons Cleaning Kit

Table 1 lists the items that are included in the weapons cleaning kit. The item number in the table corresponds with the callout number in the illustration.

Table 1. Weapons Cleaning Kit.

ITEM	PART NO.	DESCRIPTION	QTY
1	RW-316	Cleaning brush	1
2	RW-970	Panoply patches	10
3	RW-CLP-16	2/3 ounces Breakfree	1
4	IP-9098-5	Extension rod	1
5	IP-557-S-SCRW	Cleaning eye patch rod	1
6	IP-555-5	T-handle obstruction remover	1
7	IP-C-40-50-2	Cleaning cable	1
8	558-B-3-132	Bronze bore brush	1
9	7790737	Bronze chamber brush	1
10	82135-B	Case, maintenance equipment, small arms	1

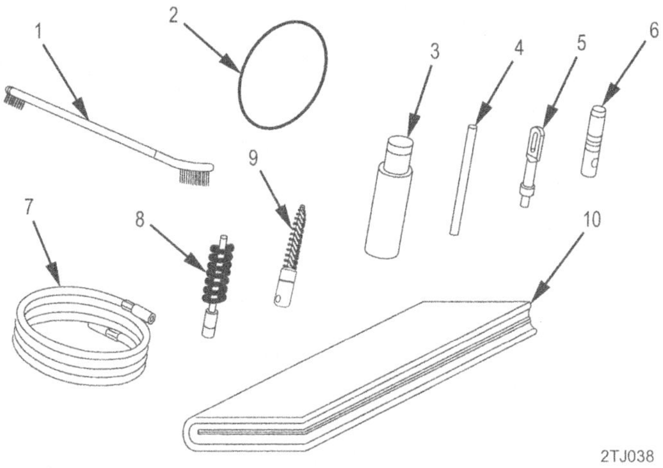

TM 9-1005-239-10 0018 00

DEPLOYMENT KIT – Continued

Optical Cleaning Kit

Table 2 lists the items that are included in the optical cleaning kit. The item number in the table corresponds with the callout number in the illustration.

Table 2. Optical Cleaning Kit.

ITEM	PART NO.	DESCRIPTION	QTY
1	A-A-3191	Brush, artist (cleaning brush)	1
2	96041	Case, lens cleaning	1 ea
3	L-B-56A	Bottle (for containing isopropyl alcohol)	1
4	NNN-P-40	Paper, lens (cleaning tissue)	1 pkg
5	362	Applicator (Q-tips) 100 per pack	1 bx

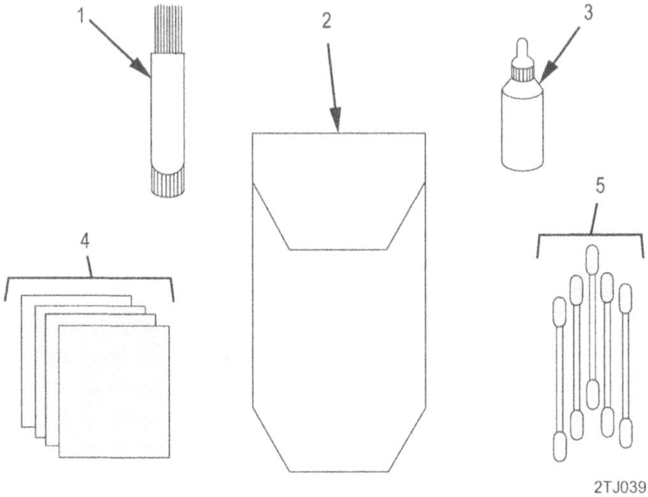

DEPLOYMENT KIT – Continued

Deployment Tool Kit

Table 3 lists the items that are included in the deployment tool kit. The item number in the table corresponds with the callout number in the illustration.

Table 3. Deployment Tool Kit.

ITEM	PART NO.	DESCRIPTION	QTY
1	96059	65 lb/in torque wrench (3/8 drive) w/socket 1/2 in socket (3/8 drive) (for use with torque wrench): B107.1 CL1STA	1
2	1162	1/2 in. short handle combination wrench	1
3	T-33	1/16 roll pin punch:	1
4	T-5	T30 L-shaped Torx® wrench	1
5	T-46	T15 L-shaped Torx® wrench	1
6	T-45	T10 L-shaped Torx® wrench	1
7	T-47	0.050 L-shaped allen wrench	1
8	T-51	3/32 in. L-shaped allen wrench	1

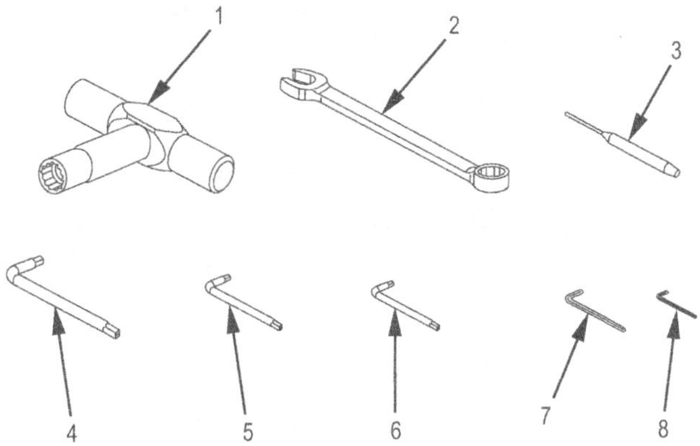

DEPLOYMENT KIT – Continued

Deployment Parts Kit

Table 4 lists the items that are included in the deployment parts kit. The item number in the table corresponds with the callout number in the illustration.

Table 4. Deployment Parts Kit.

ITEM	PART NO.	DESCRIPTION	QTY
1	13EYE	Cover, eyepiece lens	1
2	403BJ	Cover, objective lens	1
3	82114-A	Bipod or Rear Lock Pin	1
4	82115-A	Midlock pin	1
5	51503	Cover, turret	1
6	82125	Cap screw	2
7	82108	Plunger, extractor	1
8	82107	Spring, extractor	1
9	82106	Extractor	1
10	82128	Nut, scope ring	1
11	82129	Lock washer, scope	1
12	82127	Cap ring screw	2
13	82102	Bolt spring	1

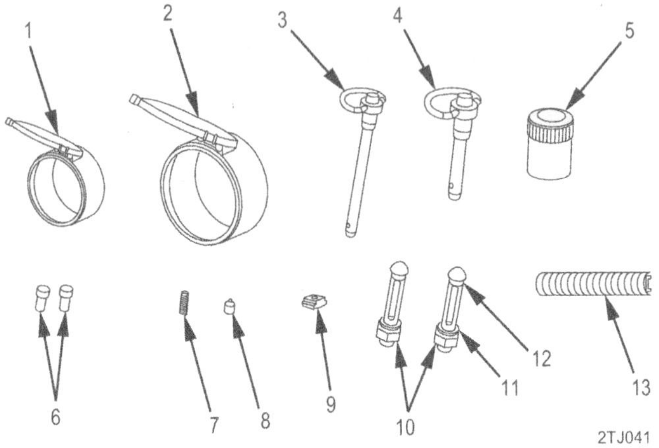

DEPLOYMENT KIT – Continued

Five Piece Cleaning Rod

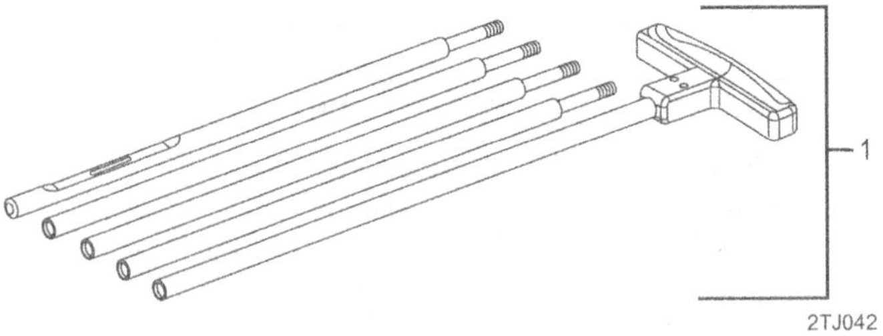

A removable handle is included with the five cleaning rods.

Table 5. Five Piece Cleaning Rod.

ITEM	PART NO.	DESCRIPTION	QTY
1	6535441	Five Piece Cleaning Rod	1

CLEANING INSTRUCTIONS

NOTE

The rifle should be cleaned and lubricated as soon as possible after each shooting session, to prevent the corrosive effects of moisture, and buildup of debris, grit, and carbon build up in the action and barrel.

Cleaning the Chamber

1. Using deployment cleaning kit (item 8, WP 0021 00), insert a section of the five piece cleaning rod through the ratchet handle of the chamber brush. Apply cleaning solvent to the brush and vigorously scrub the chamber. Turn the cable and brush in a clockwise direction in order to remove residue from the corner of the neck.

CLEANING INSTRUCTIONS - Continued

Cleaning the Optics

CAUTION

Use extreme care to avoid scratching the lenses. Do not allow solvents to touch the lenses.

1. Remove dust, lint, and dirt from the lens and exterior of the scope using a clean camel hair artist brush.

2. To remove smudges from the lens, breathe heavily on them and wipe off moisture with lens paper. If lens paper is not available, use a soft, clean, dry cloth.

3. Keep all hex screw fittings clear of mud and dirt. If they become clogged, use a safety pin or similar item to dig out the debris.

4. Keep lens free of oil and grease. If there are fingerprints, oil spots, etc., on the lens, use rubbing alcohol with lens paper to remove. Pat the lens, do not scrub.

5. After cleaning, apply a light coat of lubricant to the scope body.

Cleaning the Muzzle and Muzzle Brake

1. On one end of the cleaning cable attach the T-handle body and insert the rod handle through the hole in the T- handle body. Pass the opposite end of the cable through the muzzle end of the barrel until the end exits the chamber.

2. Attach the cleaning tip with two patches dipped in cleaning solvent, and pull the patches out through the muzzle to remove powder residue. Repeat with the bore brush dipped in solvent. Attach two clean patches to the cleaning tip and repeatedly pull clean patches out through the muzzle end until they come out clean.

3. Clean the muzzle brake with a small brush and bore solvent. It is best to clean the muzzle brake at the same time the barrel is being cleaned, as the bore solvent will help in loosening the carbon that builds-up on the interior walls.

CLEANING INSTRUCTIONS - Continued

Cleaning the Bolt Assembly

Clean the bolt face with bore solvent. Use a brush and scraper to remove carbon and brass shavings from both the extractor and the ejector. Depress the ejector and extractor by hand to test their smooth motion. If they hang-up or their motion is not smooth, remove them and clean the parts, springs and holes. Apply lubrication before assembly, and test their motion by hand.

Cleaning the Magazine

WARNING

Wear eye protection to prevent injury from spring-loaded parts, particularly when removing the magazine base plate.

NOTE

Disassembling the magazine is not recommended as a matter of routine maintenance, but it may become necessary for repairs.

1. Hold magazine tube upside down on a firm surface and place end of 1/8 in. punch in hole located on base plate. Gently pry upwards to clear locking flange on base plate, and slide base plate off magazine tube. (It may be necessary to tap base plate a few times with punch to get it started).
2. Control magazine spring as base plate is removed. Withdraw magazine spring and magazine follower from magazine tube.
3. Wipe clean all parts of magazines.

WARNING

Wear eye protection to prevent injury from spring-loaded parts, particularly when replacing the magazine base plate.

4. Ensure that loop of magazine spring is around protrusion located on bottom of magazine follower.
5. Install magazine follower and magazine spring into magazine tube. Secure with base plate.

CLEANING INSTRUCTIONS - Continued

6. After cartridge magazine has been reassembled, check for proper functioning by loading it with five dummy rounds and pushing downward on the dummy rounds. They should move freely without binding.

END OF WORK PACKAGE

CHAPTER 5

AMMUNITION MAINTENANCE INSTRUCTIONS FOR LONG RANGE SNIPER RIFLE, M107

OPERATOR
LONG RANGE SNIPER RIFLE, M107
NSN 1005-01-469-2133
CARE, HANDLING, AND PRESERVATION OF AMMUNITION

GENERAL INFORMATION

WARNING

DO NOT FIRE seriously corroded ammunition, dented cartridges, cartridges with loose bullets, cartridges exposed to extreme heat (135°F (57.2°C)) until they have cooled, or cartridges with pushed in (short) bullets.

The use of unauthorized ammunition may cause serious injury to personnel and damage to the rifle.

CAUTION

Check ammunition and magazine frequently for corrosion. If necessary, clean ammunition with a dry cloth. Do **NOT** lubricate ammunition.

GENERAL INFORMATION - Continued

CAUTION - Continued

Do not open ammunition containers until the ammunition is to be used. Ammunition removed from the airtight containers, particularly in damp climates, is likely to corrode.

Protect ammunition from mud, dirt, and water. If the ammunition gets wet or dirty, wipe it off prior to use. Wipe off light corrosion as soon as it is discovered. Heavily corroded cartridges or cartridges which have dented cases or loose projectiles should not be fired.

Do not expose ammunition to the direct rays of the sun. If the powder is hot, excessive pressure may develop when the gun is fired.

Do not oil or grease the ammunition. Dust and other abrasives collecting on oiled or greased ammunition will damage the operating parts of the gun. Oiled cartridges will produce excessive chamber pressure.

If ammunition is wet or dirty, wipe it off with a dry rag prior to use. Do not lubricate.

COMPATIBLE AMMUNITION

WARNING

Use of unauthorized ammunition in the M107 Sniper Rifle can cause personal injury or damage to rifle. Table 1 below lists the only ammunition authorized for use in the M107 Sniper Rifle. If it is not shown, it is not authorized.

Table 1. Authorized Ammunition for M107 Sniper Rifle.

Ammunition	Description
MK211 Mod 0	Caliber .50 API Cartridge
M33	Caliber .50 Ball Cartridge
M17	Caliber .50 Tracer Cartridge
M8	Caliber .50 API Cartridge
M20	Caliber .50 APIT Cartridge
M1A1	Caliber .50 Blank Cartridge

COMPATIBLE AMMUNITION - Continued

MK211 MOD 0, Caliber .50 API Cartridge

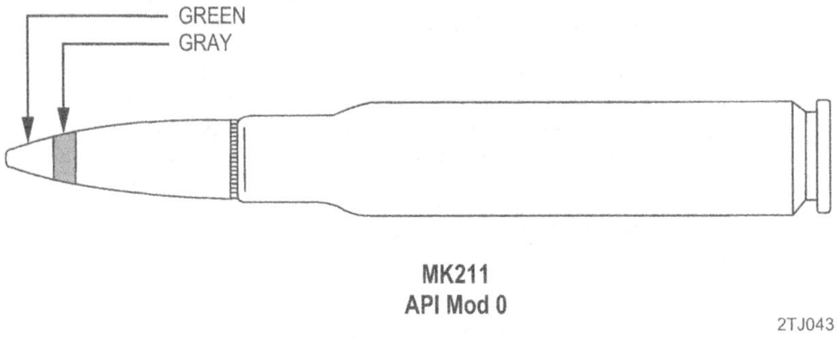

Note: Armor-Piercing Incendiary (API)

M33, Caliber .50 Ball Cartridge

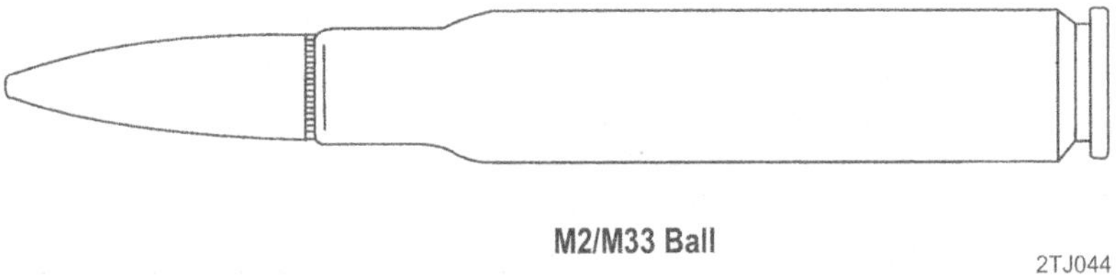

M17, Caliber .50 Tracer Cartridge

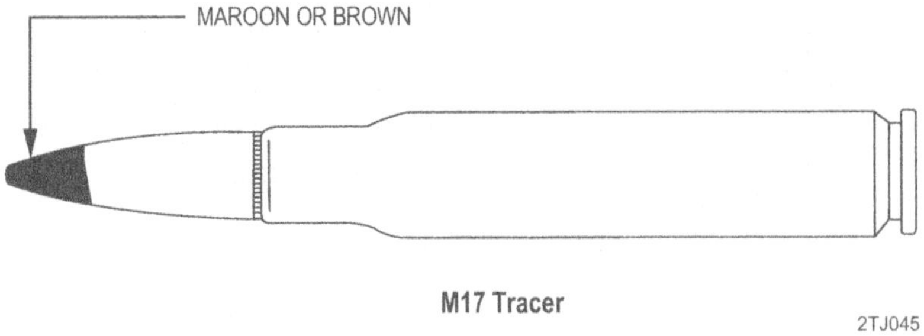

COMPATIBLE AMMUNITION - Continued

M8, Caliber .50 API Cartridge

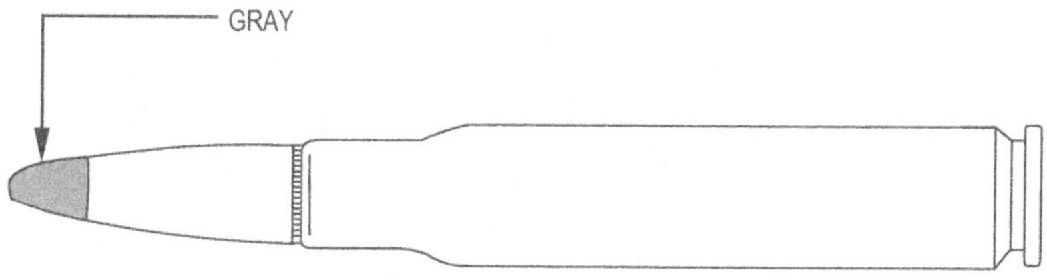

M8 API

2TJ046

M20, Caliber .50 Armor-Piercing Incendiary Tracer (APIT) Cartridge

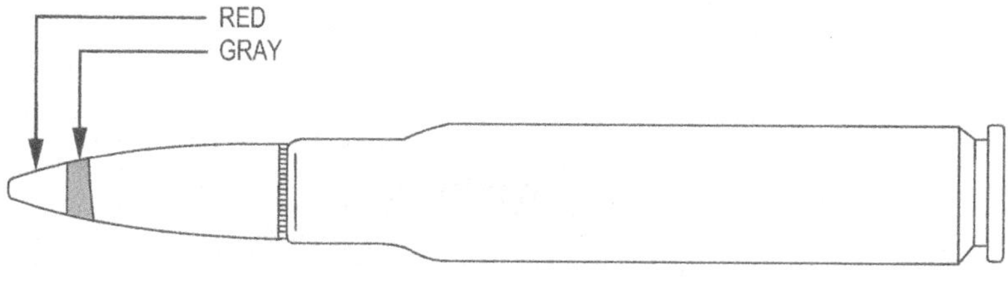

M20 APIT

2TJ047

COMPATIBLE AMMUNITION - Continued

M1A1, Caliber .50 Blank

M1A1 Blank

2TJ048

END OF WORK PACKAGE

CHAPTER 6

SUPPORTING INFORMATION

OPERATOR
LONG RANGE SNIPER RIFLE, M107
NSN 1005-01-469-2133
REFERENCES

SCOPE

This work package lists all field manuals, forms, technical manuals and miscellaneous publications referenced in this manual.

FIELD MANUALS

FM 21-11	First Aid for Soldiers
FM 31-70	Basic Cold Weather Manual
FM 31-71	Northern Operations
FM 90-3	Desert Operations

MISCELLANEOUS PUBLICATIONS

MIL-STD-129	Marking for Shipment and Storage

TECHNICAL MANUALS

TM 38-750	The Army Maintenance Management System (TAMMS)
TM 750-244-7	Procedures for Destruction of Equipment to Prevent Enemy Use

END OF WORK PACKAGE

OPERATOR
LONG RANGE SNIPER RIFLE, M107
NSN 1005-01-469-2133
SNIPER WEAPONS SYSTEM PARTS LIST

INTRODUCTION

Explanation of Columns in the Sniper Weapons System Parts List

Column (1) Illus Number. Gives you the number of the item illustrated.

Column (2) SMR Code. The SMR code contains supply/requisitioning information, maintenance level authorization criteria, and disposition instructions.

Column (3) National Stock Number (NSN). Identifies the stock number of the items to be used for requisitioning purposes.

Column (4) Description, CAGEC, and Part Number. Identifies the Federal item name (in all capital letters) followed by a minimum description when needed. The last line below the description is the CAGEC (Commercial and Government Entity Code) (in parentheses) and the part number.

Column (5) Qty. Indicates the quantity required.

SNIPER WEAPONS SYSTEM PARTS LIST

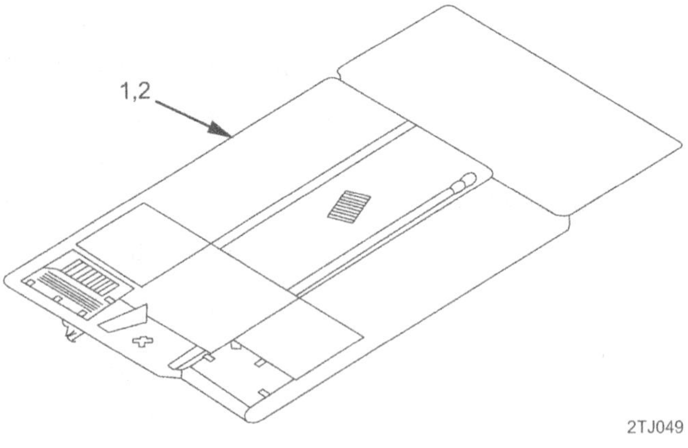

Figure 1. Drag Bag, Unfolded, 82143-3

Table 1. M107 Sniper Rifle Parts List.

(1) ITEM NO.	(2) SMR CODE	(3) NATIONAL STOCK NUMBER	(4) DESCRIPTION, CAGEC, PART NUMBER	(5) QTY
1	XAOZZ		FIG. 1 DRAG BAG, UNFOLDED (flat) (OBT64) 82143-3	1
2	PAOOO		BAG, ORDINANCE, WEAPON (OBT64) 82143-3A	1

SNIPER WEAPONS SYSTEM PARTS LIST - Continued

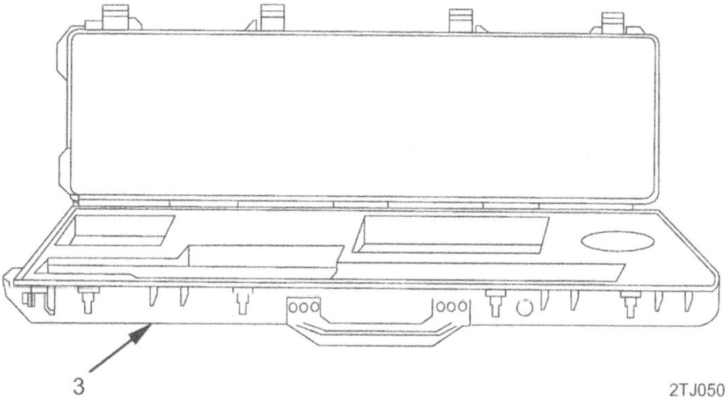

Figure 2. Hard Case, 82133-1

Table 1. M107 Sniper Rifle Parts List - Continued.

(1) ITEM NO.	(2) SMR CODE	(3) NATIONAL STOCK NUMBER	(4) DESCRIPTION, CAGEC, PART NUMBER	(5) QTY
3	PAOZZ	1005-01-415-0154	FIG. 2 HARD CASE (OBT64) 82133-1	1

SNIPER WEAPONS SYSTEM PARTS LIST - Continued

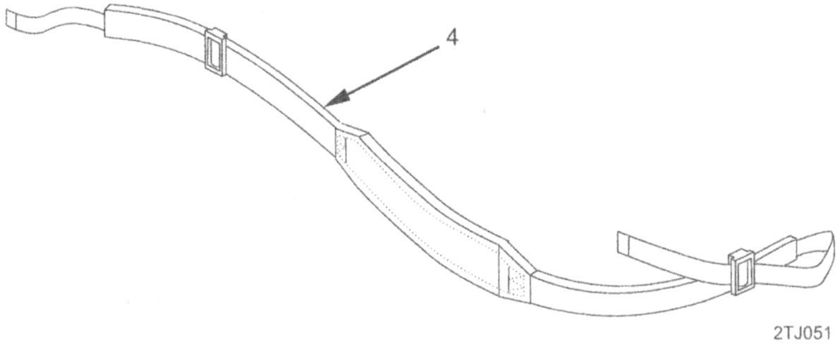

Figure 3. Sling, Small Arms, 12002983

Table 1. M107 Sniper Rifle Parts List - Continued.

(1) ITEM NO.	(2) SMR CODE	(3) NATIONAL STOCK NUMBER	(4) DESCRIPTION, CAGEC, PART NUMBER	(5) QTY
4	PACZZ	1005-00-312-7177	FIG. 3 SLING, SMALL ARMS (19204) 12002983	1

SNIPER WEAPONS SYSTEM PARTS LIST - Continued

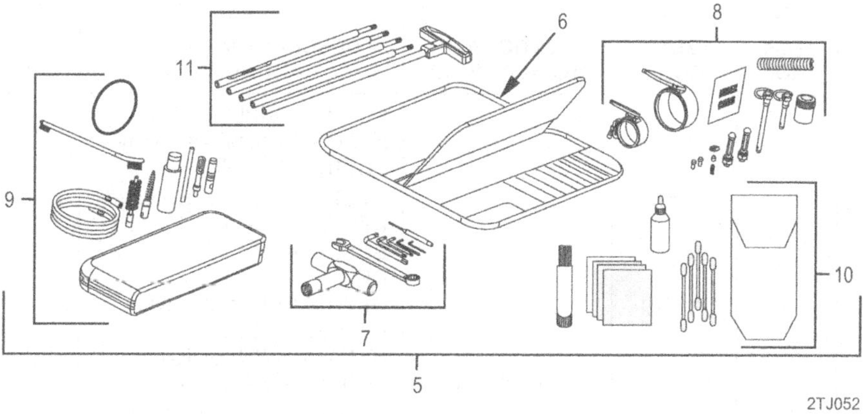

Figure 4. Deployment Kit, DK-1

Table 1. M107 Sniper Rifle Parts List - Continued.

(1) ITEM NO.	(2) SMR CODE	(3) NATIONAL STOCK NUMBER	(4) DESCRIPTION, CAGEC, PART NUMBER	(5) QTY
5	AOOOO		FIG. 4 DEPLOYMENT KIT (INCLUDES ITEMS 6-11) DK-1	
6	PACZZ	5140-01-512-8449	DEPLOYMENT KIT BAG (OBT64) DK-B-1	1
7	AOOOO		DEPLOYMENT TOOL KIT (SEE FIG. 6 FOR KIT BREAKDOWN) TK-1	1
8	AOOOO		DEPLOYMENT PARTS KIT (SEE FIG.5 FOR KIT BREAKDOWN) PK-1	1

Table 1. M107 Sniper Rifle Parts List - Continued.

(1) ITEM NO.	(2) SMR CODE	(3) NATIONAL STOCK NUMBER	(4) DESCRIPTION, CAGEC, PART NUMBER	(5) QTY
			FIG. 4 DEPLOYMENT KIT (INCLUDES ITEMS 6-11) DK-1 (CONT)	
9	AOOOO		DEPLOYMENT CLEANING KIT (SEE FIG. 7 FOR KIT BREAKDOWN) CK-1	1
10	AOOOO		OPTICS CLEANING KIT (SEE FIG. 8 FOR KIT BREAKDOWN) OCK-1	1
11	PAOOO	1005-00-653-5441	FIVE PIECE CLEANING ROD (SEE FIG. 9 FOR KIT BREAKDOWN) 6535441	1

THIS PAGE INTENTIONALLY LEFT BLANK

SNIPER WEAPONS SYSTEM PARTS LIST - Continued

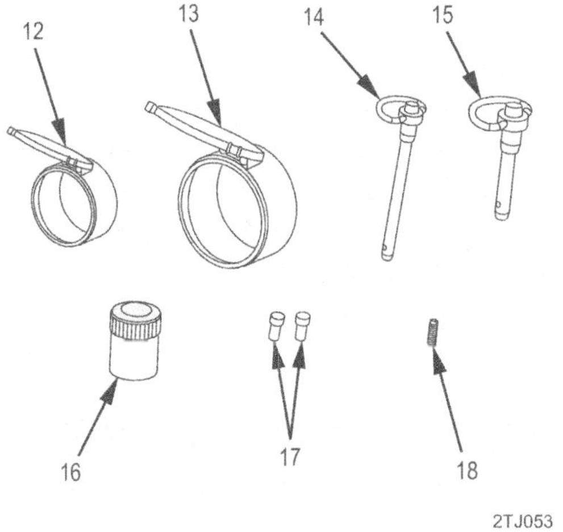

Figure 5. Deployment Parts Kit, PK-1 (Sheet 1 of 2)

Table 1. M107 Sniper Rifle Parts List - Continued.

(1) ITEM NO.	(2) SMR CODE	(3) NATIONAL STOCK NUMBER	(4) DESCRIPTION, CAGEC, PART NUMBER	(5) QTY
			FIG. 5 DEPLOYMENT PARTS KIT PK-1	
12	PACZZ	6650-01-502-1874	COVER, EYEPIECE LENS (66575) 13EYE	1
13	PACZZ	6650-01-502-1871	COVER, OBJECTIVE LENS (66575) 430BJ	1
14	PACZZ	5315-01-300-2640	PIN, QUICK RELEASE (OBT64) 82114-1A	2
15	PACZZ	5315-01-210-0923	PIN, QUICK RELEASE (OBT64) 82115-1A	1
16	PACZZ	5340-01-502-6888	COVER, TURRET (35848) 51503	1
17	PACZZ	5305-01-502-4412	SCOPE RING SCREW (35848) 53073	2
18	PACZZ	1005-01-357-4805	PLUNGER, EXTRACTOR (OBT64) 82108	1

SNIPER WEAPONS SYSTEM PARTS LIST - Continued

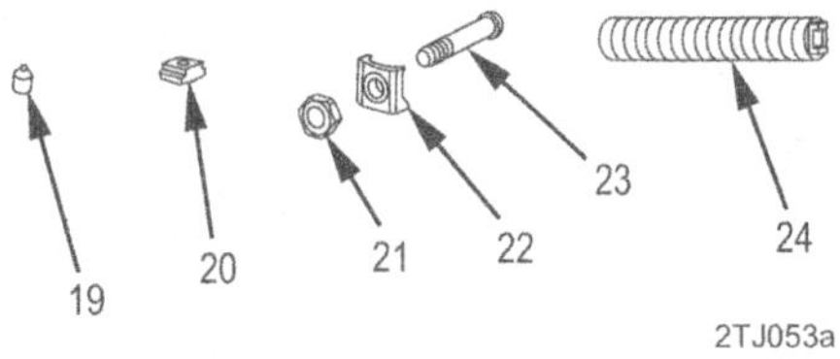

Figure 5. Deployment Parts Kit, PK-1 (Sheet 2 of 2)

Table 1. M107 Sniper Rifle Parts List - Continued.

(1) ITEM NO.	(2) SMR CODE	(3) NATIONAL STOCK NUMBER	(4) DESCRIPTION, CAGEC, PART NUMBER	(5) QTY
			FIG. 5 DEPLOYMENT PARTS KIT PK-1 (CONT)	
19	PACZZ	5360-01-358-3036	SPRING, EXTRACTOR (OBT64) 82107	1
20	PACZZ	1005-01-415-3868	EXTRACTOR (OBT64) 82106	1
21	PACZZ	5310-01-502-1527	BOLT NUT (35848) 42037	2
22	PACZZ	5310-01-502-1525	BOLT KEEPER (35848) 53076	2
23	PACZZ	5306-01-502-4414	SCOPE RING BOLT (35848) 42048	2
24	PACZZ	5360-01-358-3039	BOLT SPRING (OBT64) 82102	1

SNIPER WEAPONS SYSTEM PARTS LIST - Continued

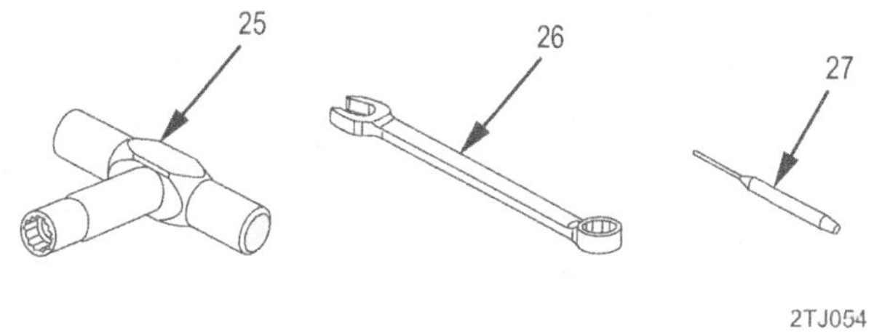

Figure 6. Deployment Tool Kit, TK-1 (Sheet 1 of 2)

Table 1. M107 Sniper Rifle Parts List - Continued.

(1) ITEM NO.	(2) SMR CODE	(3) NATIONAL STOCK NUMBER	(4) DESCRIPTION, CAGEC, PART NUMBER	(5) QTY
			FIG. 6 DEPLOYMENT TOOL KIT TK-1	
25	PACZZ	5220-01-260-2645	65 LB/IN TORQUE WRENCH (3/8 DRIVE) W/SOCKET (3A703) 96059	1
	PACZZ	5120-00-237-0977	1/2 IN. SOCKET (3/8 DRIVE)(For use with torque wrench) (05047) B107.1	1
26	PACZZ	5120-00-228-9506	1/2 IN. SHORT HANDLE COMBINATION WRENCH (96508) 1162	1
27	PACZZ	5120-01-335-1435	1/16 IN. ROLL PIN PUNCH (OBT64) T-33	1

SNIPER WEAPONS SYSTEM PARTS LIST - Continued

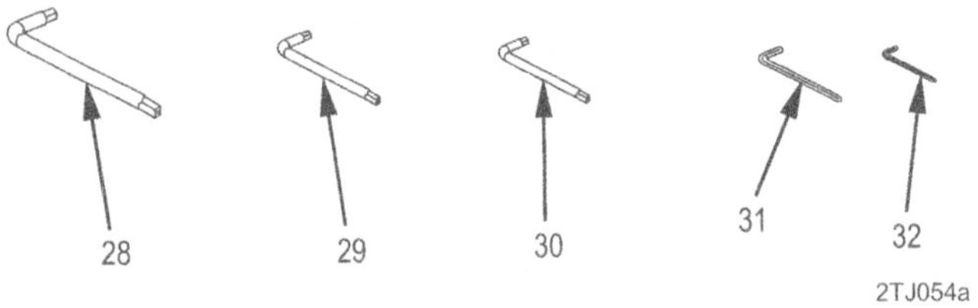

Figure 6. Deployment Tool Kit, TK-1 (Sheet 2 of 2)

Table 1. M107 Sniper Rifle Parts List - Continued.

(1) ITEM NO.	(2) SMR CODE	(3) NATIONAL STOCK NUMBER	(4) DESCRIPTION, CAGEC, PART NUMBER	(5) QTY
			FIG. 6 DEPLOYMENT TOOL KIT TK-1 (CONT)	
28	PACZZ		T30 L-SHAPED TORX® WRENCH	1
29	PACZZ		T15 L-SHAPED TORX® WRENCH	1
30	PACZZ		T10 L-SHAPED TORX® WRENCH	1
31	PACZZ	5120-00-198-5401	0.050 L-SHAPED ALLEN WRENCH (55719) AW-1-2	1
32	PACZZ	5120-00-242-7410	3/32 IN. L-SHAPED ALLEN (92674) BA27077-4	1

SNIPER WEAPONS SYSTEM PARTS LIST - Continued

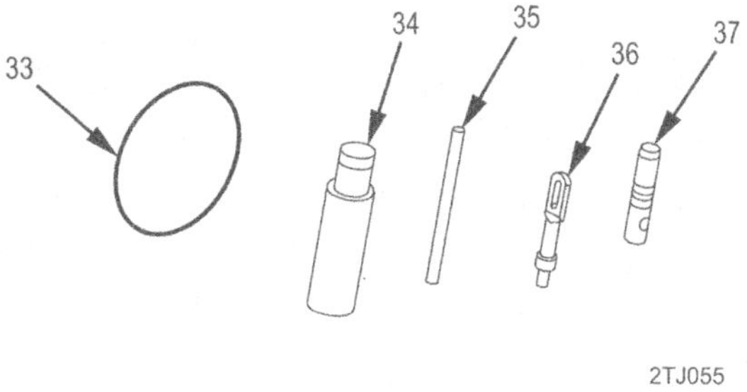

Figure 7. Deployment Cleaning Kit, CK-1 (Sheet 1 of 2)

Table 1. M107 Sniper Rifle Parts List - Continued.

(1) ITEM NO.	(2) SMR CODE	(3) NATIONAL STOCK NUMBER	(4) DESCRIPTION, CAGEC, PART NUMBER	(5) QTY
			FIG. 7 DEPLOYMENT CLEANING KIT, CK-1	
33	PACZZ	1005-01-502-5675	PANOPLY PATCHES (10 each) (OBT64) RW-970	1
34	PACZZ	9150-01-509-0426	2/3 OZ. BREAKFREE (OBT64) RW-CLP-16	1
35	PACZZ	1005-01-502-5842	EXTENSION ROD (OBT64) IP-9098-5	1
36	PACZZ	1005-01-502-5844	CLEANING EYE PATCH ROD END (OBT64) IP-557-S-SCRW	1
37	PACZZ	1005-01-512-4361	T-HANDLE OBSTRUCTION REMOVER (01VS3) IP-555-5	1

SNIPER WEAPONS SYSTEM PARTS LIST – Continued

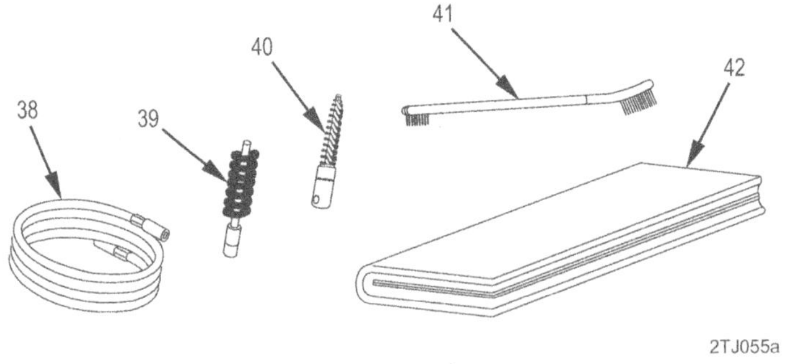

Figure 7. Deployment Cleaning Kit, CK-1 (Sheet 2 of 2)

Table 1. M107 Sniper Rifle Parts List - Continued.

(1) ITEM NO.	(2) SMR CODE	(3) NATIONAL STOCK NUMBER	(4) DESCRIPTION, CAGEC, PART NUMBER	(5) QTY
			FIG. 7 DEPLOYMENT CLEANING KIT, CK-1 (CONT)	
38	PACZZ	1005-01-502-5673	CLEANING CABLE (OBT64) IP-C-40-50-2	1
39	PACZZ	1005-01-513-4991	BRONZE BORE BRUSH (OBT64) 558-B-3-132	1
40	PACZZ	1005-00-766-0915	BRONZE CHAMBER BRUSH (19204) 7790737	1
41	PACZZ	1005-01-502-5815	CLEANING BRUSH (OBT64) RW-316	1
42	PACZZ		CASE, MAINTENANCE EQUIPMENT (OBT64) 82135-B	1

SNIPER WEAPONS SYSTEM PARTS LIST - Continued

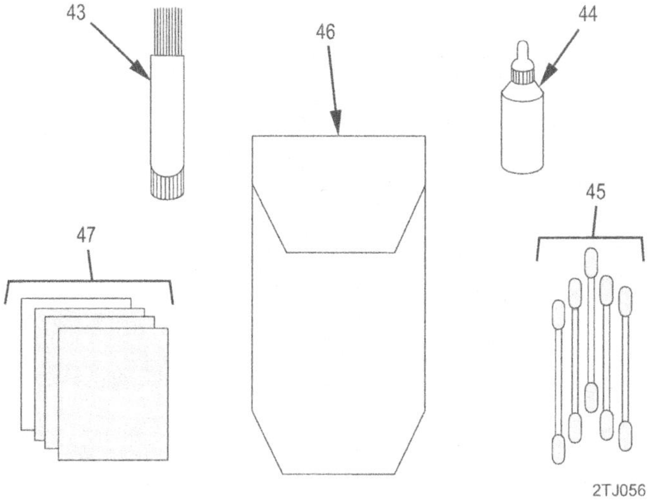

Figure 8. Optics Cleaning Kit, OCK-1

Table 1. M107 Sniper Rifle Parts List - Continued.

(1) ITEM NO.	(2) SMR CODE	(3) NATIONAL STOCK NUMBER	(4) DESCRIPTION, CAGEC, PART NUMBER	(5) QTY
			FIG. 8 Optics Cleaning Kit OCK-1	
43	PACZZ	8020-00-619-8929	BRUSH, ARTIST (cleaning brush) (58536) A-A-3191	1
44	PACZZ	8125-00-824-9058	BOTTLE (for containing isopropyl alcohol) (81348) A-A-685	1
45	PACZZ	6515-01-234-6838	APPLICATOR (Q-Tips) (5L934) 362	1
46	PACZZ	1005-01-260-2661	CASE, LENS CLEANING: (3A703) 96041	1
47	PACZZ	6640-00-663-0832	PAPER, LENS (cleaning tissue) (25518) 65-4900	1

SNIPER WEAPONS SYSTEM PARTS LIST - Continued

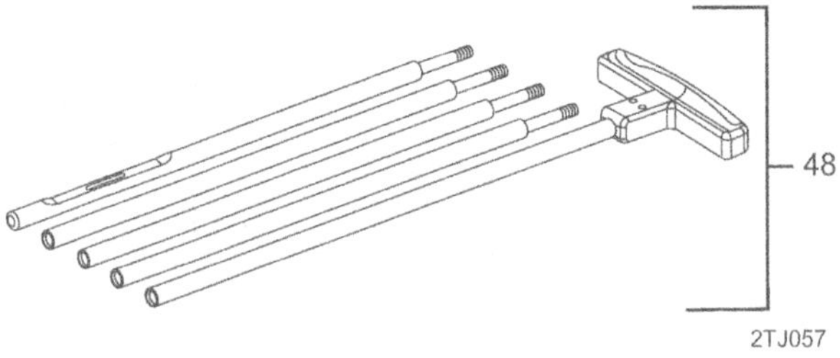

Figure 9. Five Piece Cleaning Rod, 6535441

Table 1. M107 Sniper Rifle Parts List - Continued.

(1) ITEM NO.	(2) SMR CODE	(3) NATIONAL STOCK NUMBER	(4) ITEM NAME, DESCRIPTION, CAGEC, PART NUMBER	(5) QTY
48	PAOZZ	1005-00-653-5441	FIG. 9 FIVE PIECE CLEANING ROD (19205) 6535441	1

SNIPER WEAPONS SYSTEM PARTS LIST - Continued

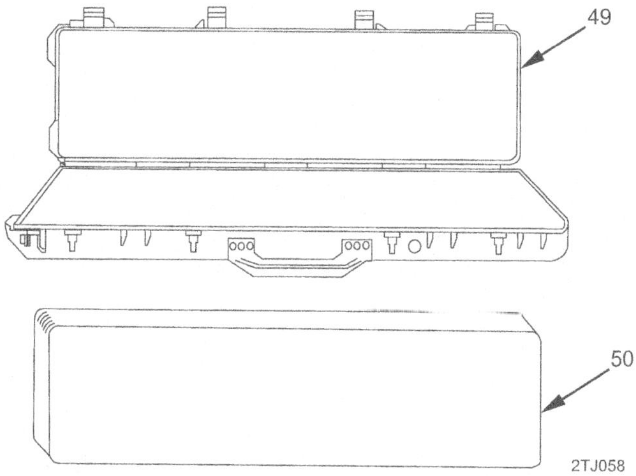

Figure 10. Auxiliary Case, 82133-1 and Auxiliary Case Insert, 82133-3-I

Table 1. M107 Sniper Rifle Parts List - Continued.

(1) ITEM NO.	(2) SMR CODE	(3) NATIONAL STOCK NUMBER	(4) DESCRIPTION, CAGEC, PART NUMBER	(5) QTY
			FIG. 10 AUXILIARY CASE, 82133-1 AND AUXILIARY CASE INSERT, 82133-3-I	
49	PAOZZ	1005-01-415-0154	AUXILIARY CASE (OBT64) 82133-1	1
50	PAOZZ		AUXILIARY CASE INSERT (OBT64) 82133-3-I	1

SNIPER WEAPONS SYSTEM PARTS LIST - Continued

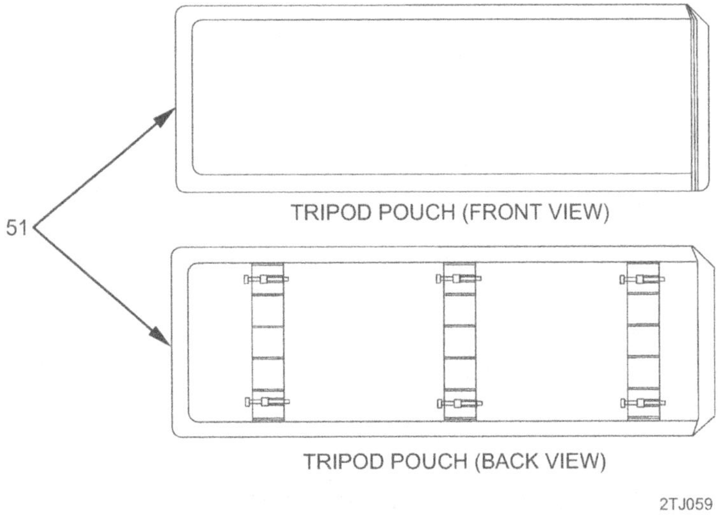

Figure 11. Equipment Weapon Bag, 82143-3-2

Table 1. M107 Sniper Rifle Parts List - Continued.

(1) ITEM NO.	(2) SMR CODE	(3) NATIONAL STOCK NUMBER	(4) DESCRIPTION, CAGEC, PART NUMBER	(5) QTY
			FIG. 11 EQUIPMENT WEAPON BAG, 82143-3-2	
51	PAOZZ	8465-01-507-5233	BAG, WEAPON, EQUIPMENT (OBT64) 82143-3-2	1

SNIPER WEAPONS SYSTEM PARTS LIST – Continued

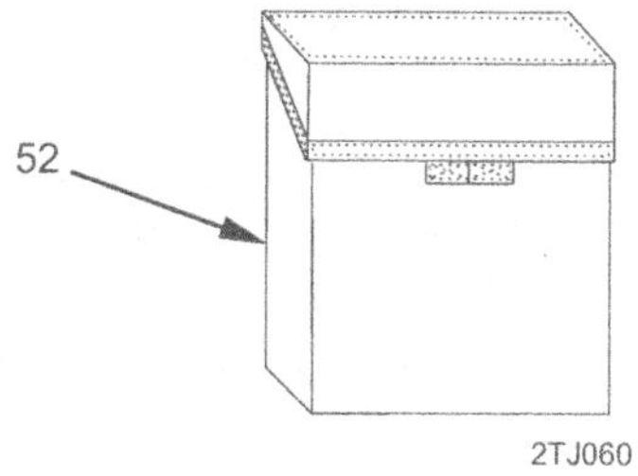

Figure 12. Optical Instrument Case, 82143-3-1

Table 1. M107 Sniper Rifle Parts List - Continued.

(1) ITEM NO.	(2) SMR CODE	(3) NATIONAL STOCK NUMBER	(4) DESCRIPTION, CAGEC, PART NUMBER	(5) QTY
			FIG. 12 OPTICAL INSTRUMENT CASE, 82143-3-1	
52	PAOZZ	1240-01-502-9044	CASE, OPTICAL INSTRUMENT (OBT64) 82143-3-1	1

SNIPER WEAPONS SYSTEM PARTS LIST – Continued

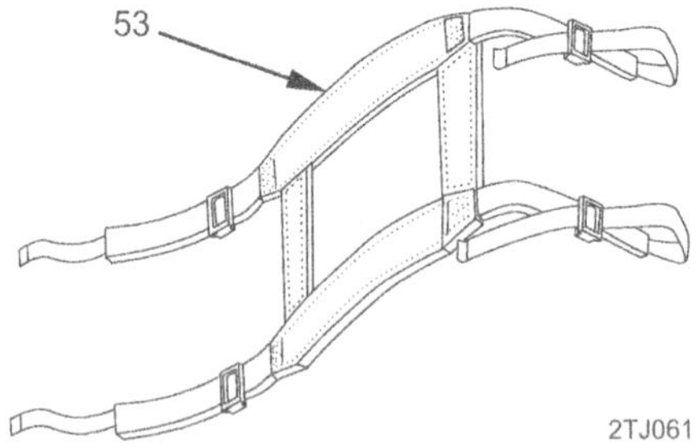

Figure 13. Shoulder Straps, 82143-3-3

Table 1. M107 Sniper Rifle Parts List - Continued.

(1) ITEM NO.	(2) SMR CODE	(3) NATIONAL STOCK NUMBER	(4) DESCRIPTION, CAGEC, PART NUMBER	(5) QTY
			FIG. 13 SHOULDER STRAPS, 82143-3-3	
53	PAOZZ	8465-01-514-8504	SHOULDER STRAPS (OBT64) 82143-3-3	1

SNIPER WEAPONS SYSTEM PARTS LIST - Continued

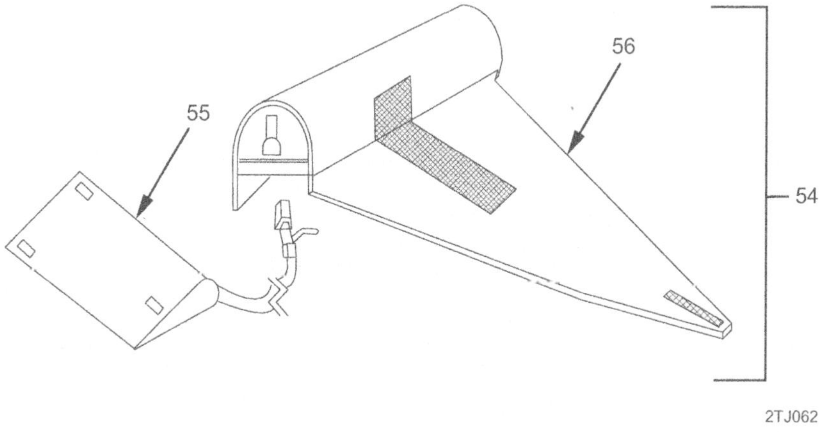

Figure 14. Scope/Muzzle Cover Assembly

Table 1. M107 Sniper Rifle Parts List - Continued.

(1) ITEM NO.	(2) SMR CODE	(3) NATIONAL STOCK NUMBER	(4) DESCRIPTION, CAGEC, PART NUMBER	(5) QTY
			FIG. 14 SCOPE/MUZZLE COVER ASSEMBLY	
54	PAOOO	1005-01-505-9110	SCOPE/MUZZLE COVER ASSEMBLY (OBT64) CVRA	1
55	PAOZZ	1005-01-504-3429	MUZZLE COVER (OBT64) 82159CVR	1
56	XAOZZ		SCOPE COVER (OBT64) 82123CVR	1

SNIPER WEAPONS SYSTEM PARTS LIST - Continued

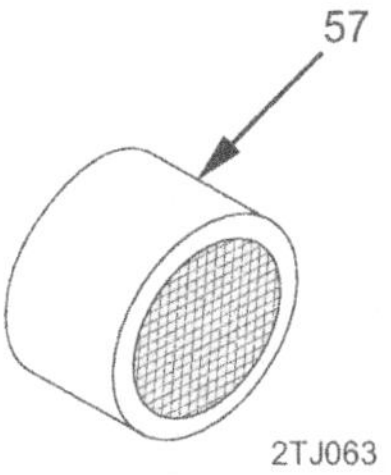

Figure 15. Anti Reflective Device (ARD)

Table 1. M107 Sniper Rifle Parts List - Continued.

(1) ITEM NO.	(2) SMR CODE	(3) NATIONAL STOCK NUMBER	(4) DESCRIPTION, CAGEC, PART NUMBER	(5) QTY
			FIG. 15 ANTI REFLECTIVE DEVICE (ARD)	
57	PAOZZ	6650-01-502-1873	ANTI REFLECTIVE DEVICE (ARD) (1D2P7) 50LTC-ARD	1

SNIPER WEAPONS SYSTEM PARTS LIST - Continued

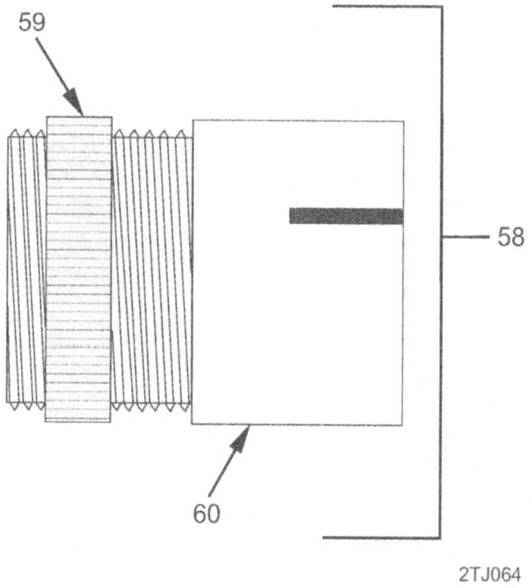

Figure 16. Laser Filter Unit (LRU), 13001589

Table 1. M107 Sniper Rifle Parts List - Continued.

(1) ITEM NO.	(2) SMR CODE	(3) NATIONAL STOCK NUMBER	(4) DESCRIPTION, CAGEC, PART NUMBER	(5) QTY
			FIG. 16 LASER FILTER UNIT (LRU), 13001589	
58	PAOOO	1240-01-502-1295	LASER FILTER UNIT (LRU) (19200) 13001589	1
59	PAOZZ	5310-01-502-1943	KNURLED LOCK RING (19200) 13001590	1
60	XAOZZ		LASER FILTER CELL (19200) 13001591	1

SNIPER WEAPONS SYSTEM PARTS LIST - Continued

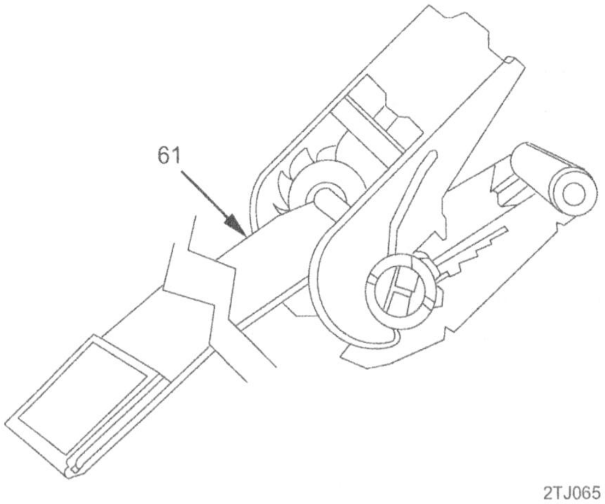

Figure 17. Carrying Case Strap, 82133-ST

Table 1. M107 Sniper Rifle Parts List - Continued.

(1) ITEM NO.	(2) SMR CODE	(3) NATIONAL STOCK NUMBER	(4) DESCRIPTION, CAGEC, PART NUMBER	(5) QTY
			FIG. 17 CARRYING CASE STRAP, 82133-ST	
61	PAOZZ	5340-01-504-8516	CARRYING CASE STRAP (OBT64) 82133-ST	2

SNIPER WEAPONS SYSTEM PARTS LIST - Continued

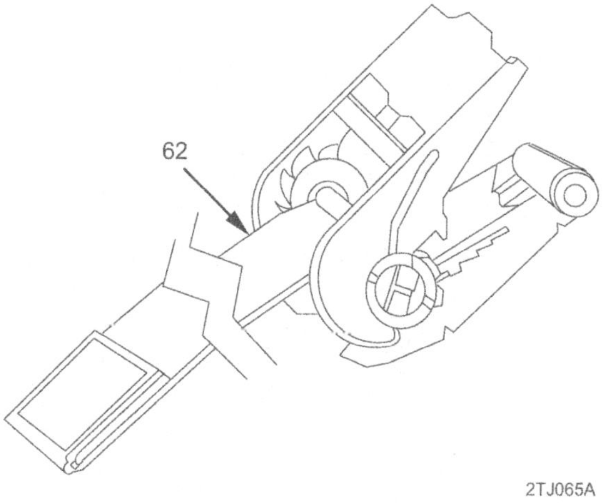

Figure 18. Carrying Case Strap, 82133-ST1

Table 1. M107 Sniper Rifle Parts List - Continued.

(1) ITEM NO.	(2) SMR CODE	(3) NATIONAL STOCK NUMBER	(4) DESCRIPTION, CAGEC, PART NUMBER	(5) QTY
			FIG. 18 CARRYING CASE STRAP, 82133-ST1	
62	PAOZZ	5340-01-504-8513	CARRYING CASE STRAP (OBT64) 82133-ST1	1

SNIPER WEAPONS SYSTEM PARTS LIST - Continued

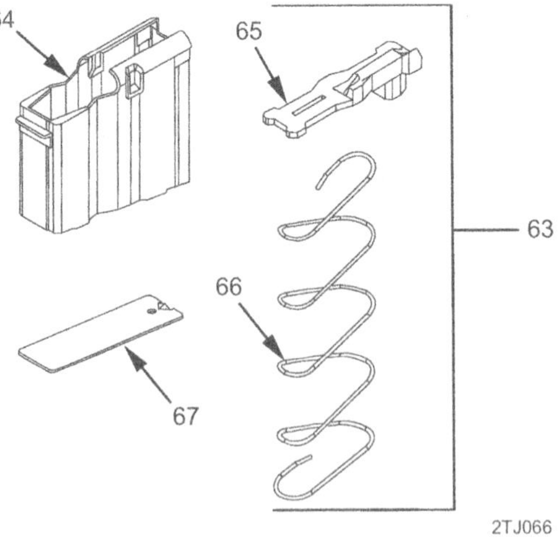

Figure 19. Cartridge Magazine, 82116-A

Table 1. M107 Sniper Rifle Parts List - Continued.

(1) ITEM NO.	(2) SMR CODE	(3) NATIONAL STOCK NUMBER	(4) DESCRIPTION, CAGEC, PART NUMBER	(5) QTY
			FIG. 19 CARTRIDGE MAGAZINE	
63	PAOZZ	1005-01-358-1342	MAGAZINE, CARTRIDGE (82116A) 82116-A	6
64	XAOZZ		MAGAZINE TUBE (OBT 64) 82116-C	1
65	PAOZZ	1005-01-358-9718	MAGAZINE FOLLOWER (OBT 64) 82120	1
66	PAOZZ	5630-01-358-7910	MAGAZINE SPRING (OBT 64) 82121	1
67	PAOZZ	1005-01-358-9712	MAGAZINE FLOOR PLATE (OBT 64) 82122	1

END OF WORK PACKAGE

OPERATOR

LONG RANGE SNIPER RIFLE, M107

NSN 1005-01-469-2133

EXPENDABLE AND DURABLE ITEMS LIST

INTRODUCTION

Scope

This work package lists expendable and durable items that you will need to operate and maintain the M107 Sniper Rifle. This list is for information only and is not authority to requisition the listed items. These items are authorized to you by CTA 50-970, Expendable/Durable Items (Except Medical, Class V, Repair Parts, and Heraldic Items), or CTA 8-100, Army Medical Department Expendable/Durable Items.

Explanation of Columns in the Expendable/Durable Items List

Column (1) Item Number. This number is assigned to the entry in the list and is referenced in the narrative instructions to identify the item (e.g., Use anti-fogging compound (item 3, WP 0022 00)).

Column (2) Level. This column identifies the lowest level of maintenance that requires the listed item (C = Operator/Crew).

INTRODUCTION - Continued

Column (3) National Stock Number (NSN). This is the NSN assigned to the item which you can use to requisition it.

Column (4) Item Name, Description, Commercial and Government Entity Code (CAGEC), and Part Number (P/N). This column provides the other information you need to identify the item.

Column (5) Unit of Measure (U/M). Indicates the physical measurement or count of the item as issued per the National Stock Number shown in column (3).

EXPENDABLE AND DURABLE ITEMS LIST

Table 1. Expendable and Durable Items.

(1) ITEM NO.	(2) LEVEL	(3) NATIONAL STOCK NUMBER	(4) ITEM NAME, DESCRIPTION, CAGEC, PART NUMBER	(5) U/M
1	C	6810-00-983-8551	ALCOHOL, ISOPROPYL (cleaning fluid) 1 qt can (81348) TTI 735	QT
2	O	6515-01-234-6838	APPLICATOR (Q-tips), 100 per package (5L934) 362	PK
3	C	5120-00-254-4612	BLOWER, WATCHMAKERS (19200) 8284021	EA
4	C	8125-00-824-9058	BOTTLE (for containing isopropyl alcohol), 1 oz bottle (81348) A-A-685	OZ

EXPENDABLE AND DURABLE ITEMS LIST - Continued

Table 1. Expendable and Durable Items - Continued.

(1) ITEM NO.	(2) LEVEL	(3) NATIONAL STOCK NUMBER	(4) ITEM NAME, DESCRIPTION, CAGEC, PART NUMBER	(5) U/M
5	C	8020-00-619-8929	BRUSH, ARTIST (cleaning) (58536) A-A-3191	EA
6	C	1005-00-766-0915	BRUSH, CLEANING, SMALL ARMS (chamber) (19024) 7790737	EA
7	C	1005-00-550-4037	BRUSH, CLEANING, SMALL ARMS (M4 bore) (19204) 5504037	EA
8	O	9150-01-102-1473	CLEANER, LUBRICANT AND PRESERVATIVE (CLP), 1/2 oz bottle (83149) MIL-PRF-63460	OZ

Table 1. Expendable and Durable Items - Continued.

(1) ITEM NO.	(2) LEVEL	(3) NATIONAL STOCK NUMBER	(4) ITEM NAME, DESCRIPTION, CAGEC, PART NUMBER	(5) U/M
9	C	9920-00-292-9946	CLEANER, TOBACCO PIPE (89855) DILLSPIPECLEANERS	BX
10	O	6850-00-392-9751	CLEANING COMPOUND, OPTICAL LENS, 2 oz (58536) A-A-59199	OZ
11	C		CLEANING COMPOUND, RIFLE BORE COMPOUND (RBC) SOLVENT	OZ
		6850-00-224-6656	2 oz can (81349) MIL-PRF-372	
		6850-00-224-6657	8 oz can (81349) MIL-PRF-372	
12	O	6850-00-754-2671	COMPOUND, ANTI-FOGGING (62639) ALFA KLEEN AK-03	OZ

EXPENDABLE AND DURABLE ITEMS LIST - Continued

Table 1. Expendable and Durable Items - Continued.

(1) ITEM NO.	(2) LEVEL	(3) NATIONAL STOCK NUMBER	(4) ITEM NAME, DESCRIPTION, CAGEC, PART NUMBER	(5) U/M
13	C	9150-00-292-9689	LUBRICATING OIL (LAW), 1 qt can (81349) MIL-PRF-14107	QT
14	O/C		LUBRICATING OIL (LSA)	OZ
		9150-00-935-6597	2 oz plastic bottle (81349) MILL46000	
		9140-00-889-3522	4 oz plastic bottle (19204) 8436793	
		9150-00-687-4241	32 oz (81349) MILL46000	
		9150-00-753-4686	1 gallon	

Table 1. Expendable and Durable Items - Continued.

(1) ITEM NO.	(2) LEVEL	(3) NATIONAL STOCK NUMBER	(4) ITEM NAME, DESCRIPTION, CAGEC, PART NUMBER	(5) U/M
15	O/C		LUBRICATING OIL (LSAT)	OZ
		9150-00-949-0323	8 oz tube (81349) MIL-L-46150	
		9150-01-109-7793	1 pound can (81349) MIL-I-46150	
16	O	6640-00-663-0832	PAPER, LENS (cleaning tissues), 50 each sheet (25518) 65-4900	BK
17	C	7920-00-205-1711	RAG, WIPING (80244) 7920-00-205-1711	LB

EXPENDABLE AND DURABLE ITEMS LIST - Continued

Table 1. Expendable and Durable Items - Continued.

(1) ITEM NO.	(2) LEVEL	(3) NATIONAL STOCK NUMBER	(4) ITEM NAME, DESCRIPTION, CAGEC, PART NUMBER	(5) U/M
18	C	1005-00-288-3565	SWAB, SMALL ARMS CLEANING (patches) (19204) 5019316	PG
19	F	7510-00-074-4952	TAPE, PRESSURE SENSITIVE, cloth-back, water-resistant, 2 inches wide, 60 yard roll (81346) ASTM D 5486/D 5486M TY4 WHT 2 IN	YD

END OF WORK PACKAGE

INDEX

Subject | WP Sequence No. - Page No.

A

B

INDEX - Continued

Subject | WP Sequence No. - Page No.

B - Continued

C

Subject | WP Sequence No. - Page No.

C - Continued

D

INDEX - Continued

Subject | WP Sequence No. - Page No.

D - Continued

E

Subject — WP Sequence No. - Page No.

F

INDEX - Continued

Subject — WP Sequence No. - Page No.

L

Subject | WP Sequence No. - Page No.

M - Continued

N

INDEX - Continued

Subject | WP Sequence No. - Page No.

O

Subject | WP Sequence No. - Page No.

INDEX - Continued

Subject | WP Sequence No. - Page No.

Made in the USA
Monee, IL
04 November 2026

67384031R00066